Text Me IF YOU'RE BREATHING

Observations, Frustrations, and Life Lessons from a Low-Tech Dad

GREG SCHWEM

www.comedywithabyte.com

First Printing: July 2010

Cover photography by Paul Crisanti
Cover and interior design by Judine O'Shea

ISBN-10 061537624X
ISBN-13 9780615376240

Library of Congress Cataloguing-in-Publication Data
has been applied for. Available upon request

Printed in the United States of America
9 8 7 6 5 4 3 2 1

Text Me If You're Breathing

Observations, Frustrations, and Life Lessons from a Low-Tech Dad

GREG SCHWEM

CONTENTS

OBSERVATIONS

CHAPTER 1

What Would Dr. Huxtable Do?

My thirteen-year-old came charging into the house as the school bus pulled away. Another day completed.

"Whoa, whoa, whoa," I said as she ran by. "What about homework? Have you unloaded the dishwasher? Practiced piano? Cleaned your room?"

Natalie replied with the only response a thirteen-year-old can muster in any situation. "I know, I know," she grumbled. "But first, I have to feed Claire."

Claire? Who was Claire? Did we have a houseguest I was unaware of? Had my wife gone "Brangelina" (the tabloid name for Brad Pitt and Angelina Jolie in case you're among the minute section of the population that doesn't get its news from *Access Hollywood*) and adopted a child without telling me?

"Excuse me," I said, as Natalie stuffed a handful of Cheez-Its in her mouth. Cheez-Its, by the way, are the official snack food of kids ages five to thirteen. The square, perforated crackers provide exactly zero nutritional value yet manage to leave a burnt orange powdery residue on every piece of furniture in the house. It's like dipping your hand into a box of rust.

"Who is Claire?"

"Claire, my pet," she replied, as if this question were so dumb, it could only have been asked by Larry King.

My brow furrowed deeper. I must be traveling too much, I thought. Apparently we acquired a pet while I was gone and nobody bothered to tell me.

As far as I knew, we hadn't had a pet since our Yorkshire terrier Barnaby went to doggie heaven. Of course I'm not counting goldfish, which Natalie and her eight-year-old sister Amy win each year at the church carnival by lobbing ping-pong balls directly into glass bowls filled with just enough water to keep a single hapless fish alive. Once my girls are declared "winners," the fish are transferred to Ziploc bags. The bags are the fish version of hospice: there is nothing more that can be done. Once home, the fish are named, given enough "fish food" to satisfy a sperm whale and succumb several hours later. Cause of death? Lack of water and overabundance of nutrition.

"I'll ask again," I said, growing slightly more agitated. "WHO is Claire?"

"My *online* pet," Natalie replied. "I didn't feed her yesterday. If I don't feed her today, she'll *DIE*!"

And with that, she rushed to the family PC, logged onto something called Webkinz.com, typed in her user name and password and was soon chatting with school friends, playing online games with names like *Wheel of Wow* and acquiring valuable points so she could buy Claire a blender. Apparently Claire's "diet" includes margaritas.

I left her in nurturing mode and went upstairs where I found Amy staring at a GameBoy, perched precisely two inches from her nose.

"Hey honey. Want to go ride bikes with Dad?"

"In a minute," she said to no one in particular. "Ariel needs to catch one more jewel."

"And Dad," she continued, "can we TiVo *Biggest Loser* tonight?"

Without replying, I retreated downstairs to our family room. The 50-inch flat-screen television hung above the fireplace; the "universal" remote that controlled the TV, DVD player, radio, iPod docking station and probably my electric toothbrush lay on the couch; the laptop with wireless Internet connectivity was humming on the kitchen table and I heard the electric garage door opening, signaling that my wife Sue was about to pull up in her Lincoln Navigator, complete with voice-activated GPS.

I glanced upstairs, knowing that my youngest was not coming down anytime soon. Once Ariel had been crowned Jewel Queen, my daughter would undoubtedly replace the cartridge with another, entitled *Dora the Explorer and the All Day Quest for the Hidden Treasures that Adults Can't Find Even If They Press Every Button on the Freaking Console!*

In no way was GameBoy Amy's sole playmate. Just the day before, she sat at the kitchen table with a seven-year-old neighbor, an inquisitive type who spent what seemed like an eternity staring at a device on my kitchen wall. She then took the Popsicle out of her mouth long enough to ask a simple and direct question:

"What's that?"

"That" was a Sony IT-ID20 *corded* telephone. I'm convinced this child knew she was staring at a phone—it was the cord that threw her.

At that moment I felt like inserting my dentures, donning a pair of wool slippers, turning up my hearing aid and saying, "Pull up a chair young whippersnapper and let Grandpa tell you what it was like during the Depression. First I had to get up at 5 a.m. and milk the cows...."

I have no idea how long we've had the Sony IT-ID20. Perhaps it was a wedding gift; maybe it was from my first apartment. The previous homeowner definitely didn't leave it behind because there was no previous owner. We built the house ourselves and chose the Sony IT-ID20 as an accessory. To Sue and I, it is a communication device; to a seven-year-old it's a mystery, an antique or both.

That phone is a reminder of a childhood my kids will never know. It seems like only yesterday that it was me sitting at the kitchen table, sucking on a Popsicle while my mother talked on a flesh-toned telephone mounted to the wall. When it rang, we immediately looked to the wall as opposed to now, when we look under couch cushions for the cell phone. We know the phone is under the cushions, we know exactly which cushion to grab and toss aside (it's the one next to the cushion where the TV remote is hiding), yet we still always manage to locate the phone a millisecond after the call has kicked to voice mail.

While my mother gabbed, I retreated to my room where I did my homework with A PENCIL. When it was completed, I listened to my collection of RECORDS, or visited friends on my BIKE because, if I wanted to speak with them, I had to actually BE IN THE SAME GENERAL AREA.

The art of parenting progressed very slowly during this time because nothing new was being invented. (Quick, name an innovation of the 1970s—the Pet Rock doesn't count.) My parents' childhoods also consisted of bicycles, records and pencils so it was easy for them to parent through experience. They had an answer for every question and never once asked for *my* advice on anything. The TV was something I needed permission to use, not something I programmed for them.

I graduated college in 1984, a year crammed with historical events, my diploma among them. On January 24, Apple Computer released the Macintosh, the very first computer that included a mouse and actual images on screen, as opposed to only text. Eight months later, *The Cosby Show* debuted on NBC. In millions of homes, mandatory Thursday night viewing included watching Bill Cosby portray Dr. Heathcliff Huxtable, an obstetrician and father to five of the most well-adjusted children in the western hemisphere. Dr. Huxtable always seemed so in touch with his kids. When they had a problem, he knew exactly how to solve it; when they requested something, he knew exactly when to adhere to their wishes and when to deny them. Why, it was almost as if he had already lived every component of their lives and therefore could parent from a handy rulebook that he kept in his desk drawer, right next to the Jell-O pudding.

Dr. Huxtable obviously did not own a Macintosh.

For that matter, neither was he forced to weigh the potential dangers of chat rooms, the addictive qualities of Facebook, or the nonverbal communication tool that is text messaging. Huxtable's conversations with his kids were always done in person. Sure he occasionally had to talk to them through a door slammed in anger but eventually that door would open, he would step inside, and proceed with a series of goofy faces that soon had his children howling, thereby letting him regain the "World's Best Dad" title before the show cut to commercial.

Incidentally, I never felt Huxtable's wife Clair campaigned for "World's Best Mom" status. She always seemed a little more distracted from family matters. Then again, she was a lawyer.

Today, parenting is more like playing sudoku, the hopelessly addictive Japanese numbers game. When you are stuck, guess. If you guess incorrectly, erase and start over. Since I don't live in Dr. Huxtable's technology-free world—and cannot make silly faces—I don't automatically have answers for my children's questions, unless one considers "huh" an answer.

Dad, the PC screen just turned blue.

Huh?

Dad, the DVD has lines running through it.

Huh?

Dad, what does "Fatal Error 231>?&4#(30)" mean?

Huh?

Dad, can we get another fish?

Huh? I mean NO, ABSOLUTELY NOT!

Just testing you, Dad.

I am also learning how to parent on the fly, something that all parents have to do since new technological innovations are being introduced, discarded and reinvented with NEW AND EXCITING FEATURES faster than the time it takes bananas to turn brown. Who cares if I'm still trying to locate the "print" button in Microsoft Word 2000? I can do that on my own time; right now I have to research the pitfalls of MySpace. I may not know how to set up the voice mail on my cell phone but that's not important. What IS important is learning how to text message on the darn thing so I can tell my daughter what time I'm picking her up. Oh, and I must do it all while in a constant state of motion since today's parents are never stationary unless they are stuck in traffic.

That's another thing I found slightly strange about Dr. Huxtable's home life; he always seemed to be *at home*. Yet he delivered babies for a living. How did that work? Was there some drug he gave patients that induced labor only between nine and five? On weekdays? If I were a stay-at-home dad like Huxtable, maybe all this kid-enticing technology wouldn't seem so foreign. Maybe I would know the names and nutritional needs of a Webkinz character; maybe I would not become alarmed when my girls say they are going down to the basement to play with "Mr. Driller W"; maybe I would know how to catch that darn jewel. Unfortunately I live in a perpetual state of cluelessness since I am not around 24/7. In 1989, I chucked my five-year career as a journalist to pursue my dream of making it as a stand-up comedian. It was a decision that thrilled nobody, least of all my parents who had paid for a Northwestern University journalism degree and were convinced my newfound profession would mean I would never make enough money to pay back tuition loans—or even the bartender at the Student Union for that matter.

Comedian, incidentally, is an interesting occupation in that it always elicits a response. You can't say the same about something like insurance salesman. Tell someone you sell insurance and they will politely mumble "mmm hmmm" before moving to the farthest corner of the room.

With comedian, the most common reaction is disbelief.

"Seriously, what do you REALLY do?"

Then there's the inevitable, "Comedian eh? Say something funny."

Comedian may be the only occupation where it is expected that you will *do* your job after identifying it. Thankfully that is not the case for every vocation. If it were, coroners would never be invited to parties.

Finally, there are the people who, once they discover you are a comedian, assume everything that comes out of your mouth is supposed to be funny. A comedian could burst into a crowded room and scream, "HELP, THERE'S A SCHOOL BUS ON FIRE AND TWENTY KIDS ARE TRAPPED INSIDE!" At least one person would chuckle and respond, "Good one!"

Shortly after I'd made my head-scratching career change, I moved from West Palm Beach, Florida, back to my hometown of Chicago—even though I was rarely there. Instead, my office consisted of a 1987 fire red Nissan Sentra. I put 100,000 miles on it in three short years, driving to gigs in such comedic hot spots as Houma, Louisiana; Bristol, Tennessee; and La Crosse, Wisconsin. The venues included comedy clubs, hotel sports bars, no-themed bars, outdoor tents bordering 18th greens, strip clubs, country clubs, military bases, concert halls, amusement parks, buses traveling seventy miles per hour down interstates and, weirdest and most humiliating of all, warm-up act at a twelve-screen movie theater!

In the last case, the owner of a Fort Lauderdale comedy club had struck a deal with the theatre manager. In exchange for placing fliers promoting the club inside the multiplex lobby, the club owner sent over comedians every Saturday night with orders to perform for five minutes in each theatre, prior to the "Coming Attractions" reel.

"It will be great exposure for you," I remember the owner telling me.

Any comedian who has been in the business more than fifteen minutes knows that "great exposure" means "disastrous consequences." I'll wager my house that the captain of the *Titanic* told the musicians to "go up on the deck and play. It will be great exposure."

Still, I was getting paid for this and since I was in the "struggling entertainer" category—a category entertainers never leave unless their name is "George Clooney"—I marched into each theatre with a handheld cordless microphone, quickly introduced myself as "Greg from Governor's Comedy Club" and proceeded with the longest five minutes of my life. Times twelve. In between mouthfuls of seven-dollar popcorn and gulps of five-dollar soda, the audience politely giggled while wondering if the projector was broken.

I have always mined material from the events currently taking place in my life. When single, I joked about cooking for one; when traveling through small town America, I mocked the tourist attractions. In the early '90s, when typewriters gave way to personal computers and electronic notebooks replaced spiral ones, I noticed the frustration level in America quickly rising, particularly in the workplace. We weren't asked if we wanted to own a computer; it was a decision made for us. Joe Office Worker came in one day to find a strange looking television screen on his desk and an accompanying keyboard. A glance around the room revealed similar contraptions on every other desk. Nobody told Joe how this stuff worked; he was expected to figure it out on his own while not missing a beat at work.

One-liner jokes in my act about installing a program, talking to an animated, on-screen paperclip and dealing with a Bangalore, India-based "help" desk led to lengthier comedic bits and eventually, an entire act based on the foibles of technology. When nightclub patrons approached me after shows and said, "You should really come down to my office and tell those jokes," it drew me into the corporate comedy field. I abandoned comedy club venues in favor of sales meetings, business retreats, product launches and awards banquets. In time, companies like Microsoft, Cisco, IBM, McDonald's and United Airlines came calling for a dose of humor courtesy of my presentation, "Comedy With a Byte." While my accomplishments fly under the radar—I've never had my own television series, appeared above a movie title or been splashed across the cover of *People* with my alleged mistress—I've done okay. My career has netted me a nice house, a quiet Midwestern lifestyle and enough free time to coach Little League, attend church, participate in Daddy Daughter Brownie Bowling and try my best to be a good husband and father. It's not always a

Norman Rockwell painting at our house, but we've made it work for sixteen years of marriage and two kids.

Ah yes, the kids. I have two children who have never used or heard of the Encyclopedia Britannica but can Google whatever they are curious about. They can locate the "delete" key but are unfamiliar with Wite-Out. They can talk to five friends in five different locations at once. They can (incorrectly) feel they are playing guitar by staring at a television screen while placing their fingers on colored buttons. They can exercise by leaping around in front of that same TV while trying to emulate dance moves from a video game called *Dance Dance Revolution.*

If their real family annoys them, they can become part of a make-believe family called *The Sims*. They can watch *American Idol* and, days later, sing into a microphone while having their skills judged by Simon, Paula and Randy; courtesy of the *American Idol* PlayStation game. In one summer afternoon, using applications like *Madden NFL, GarageBand, Final Cut Studio* and *Flight Simulator*, they can become, respectively, football coaches, musicians, movie directors or fighter pilots.

My summer afternoons consisted of tying a beach towel around my neck and pretending I was Superman.

They can download homework assignments off the Internet, e-mail their teachers with questions about what they just downloaded, and turn in assignments complete with PowerPoint. They can get a college diploma without leaving the house; attend driving school online; and RSVP to parties with a single mouse-click.

They can have hundreds of virtual friends that exist only through Facebook and MySpace. When I was growing up, virtual friends were called imaginary friends and having one meant your parents were "concerned." More than one imaginary friend earned you a trip to the school psychologist while everybody else was at gym class.

In short, my children's lives and my own are dominated by technology, the very subject that I find so amusing.

I yearn, like all parents, for more simplistic days. I want to load my daughters into the car and drive aimlessly down a country road, having a heart-to-heart talk without first having to explain why the DVD player will not be used on this trip. I want to play a family game of *Monopoly* without having to admonish them for subtly glancing at their cell phones for a text message. On that note, I also want to play *real Monopoly*, with tangible money. Parker Brothers recently introduced *Monopoly Electronic Banking Edition*, in which paper currency has been eliminated in favor of a "Banker Unit." Today's aspiring real estate tycoons start with fifteen million dollars and purchase properties by swiping their cards through the unit. In the process, they learn how easy it is to run up massive debt, a skill they will perfect in college and parlay into careers at the Federal Reserve.

The realist in me knows that the days of passing GO and collecting two hundred dollars are gone forever. It is useless to assume that the pleasures I enjoyed as a child will be duplicated in my kids' world.

Here is a perfect example: Each summer I play hooky for an afternoon and attend a major league baseball game. There are two professional teams in my backyard, the Cubs and the White Sox, and it doesn't matter which team I watch. The weather is warm,

I'm watching baseball and I get a chance to reflect on childhood memories.

Flash back to 1971, my first ball game. I attended it with the other members of Den Five, my Cub Scout troop. Thankfully we did not have to wear our Cub Scout uniforms, as there really is no reason to attend a ball game wearing knotted yellow scarves and navy pants that were always an inch too short. Instead, we dressed like the normal, wide-eyed kids we were, as we descended on Wrigley Field to watch the Chicago Cubs battle the Pittsburgh Pirates.

I brought two items from home: my baseball mitt and a ballpoint pen. Like every boy, I dreamed of snaring a foul ball from the likes of Ron Santo or Ernie Banks as if it were as routine as catching a pop fly during a Little League game. The cameras would zero in on me as the fans applauded. I'd take the ball home, put it on a bookshelf in my room and go to sleep every night staring at it.

I could dream, couldn't I?

The ballpoint pen represented a more realistic goal. I was determined to get an autograph, ANY autograph, of a major league ballplayer. I knew from watching pregame shows on TV that players liberally signed before beginning their jobs. This was my chance to not only meet one of my idols, but also come home with proof.

As Den Five made its way into Wrigley, I immediately stopped in front of the grizzled vendor yelling, "SCORECARD!" I purchased one for a dime; this was now my official autograph book as there was plenty of blank space surrounding the lineup grid for signatures. We ascended the ramp and I got my first glimpse of the ballpark. I surveyed the ivy, the manually operated scoreboard and the bleachers. Incredible sights yes, but my eyes were glued to the dugout in hopes a Cubs player would emerge and mingle with the fans.

Suddenly I spied a real live Cub. Not just any Cub but left fielder Billy Williams, my favorite player. I darted down the steps with my card, where more than fifty kids had already assembled near the third base line, all with the same intention—get Williams's signature.

For a ten-year-old boy, the scene was intimidating to say the least. Dozens of outstretched hands held programs in Williams's face. He would randomly choose one, sign and place it back into the pack, where the card's owner took it. Stretching my arm as far as I could, I did the same with my card, thinking that it was hopeless.

Moments later, I felt a tug on my scorecard, as if a fish had just nibbled my hook. I had a bite! Not just a bite but it appeared I was about to hook the mother of all fish. Through the throng of cardboard, I saw Williams's hand on MY card! I let go. As Williams signed, I wondered if I would actually get the card back. He's just going to stick it back in the pile, I thought. There are older kids here. Surely one will grab the card.

But nobody did. Williams completed his signature, placed the card back into the mass and waited for a tug. That tug came from me. I HAD THE CARD! With a whoop of delight, I retreated to my seat. It was like a simple movie premise: boy sees player, boy hands player scorecard, player signs scorecard, boy retrieves scorecard, boy lives happily ever after.

Now flash forward to the present. Picture a ten-year-old Cub Scout trying to do the same thing. How would that script be written?

Having successfully raised $800 during the winter, Den Five finally has enough money to afford baseball tickets. Or so it thinks. The den mother's plan to purchase tickets online goes awry when she downloads the Wrigley Field seat map and sees the only tickets left for the entire season are behind a support beam, never mind that it is February and Opening Day is more than two months away. A hasty bake sale/car wash/canning drive is organized and the troop raises the extra $200 needed to purchase seats through ticket broker wetookallthebestseats.com.

The boys arrive at the ballpark and endure the formalities of getting inside. Their backpacks are searched and they are wanded for explosives. One boy drifts over to purchase a scorecard even though he is breaking the den mother's first rule. On the bus ride she clearly stated, "Under no circumstances will you go ANYWHERE without adult supervision."

The boys opt not to consume any liquids, since all the chaperones are moms. They will just have to "hold it" for three hours and pray the game does not include extra innings.

Thankfully, one of the chaperones accompanies the boy to the scorecard vendor, where he completes his transaction for the simple white cardboard grid that looks exactly the same as it did thirty-nine years ago, albeit with more advertising. The other scouts decline to purchase cards, as they don't want to part with $3.50.

The youngsters take their seats and eagerly pull out seven dollars each, as the hot dog vendor has just been spotted. But the scout with the card has his eyes on the field, desperately looking for a player, any player, who is signing autographs. There are none. Yes, there are dozens of players on the field but none are talking with fans. Instead they are talking with men wearing Dolce & Gabbana suits and carrying two cell phones apiece.

Speaking of cell phones, the boys get a lesson in swearing when they are forced to listen to a spectator sitting directly behind them in company-purchased seats. He is having a loud conversation with somebody who is named either "Dickhead" or "Bill." A disapproving glance from the den mother proves pointless.

Suddenly a lone player wanders over to the stands, Sharpie in hand. The scout has no idea who it is. Truth be known, it's a minor leaguer who was called up yesterday for a ten-day assignment. But all the kid knows is that a professional baseball player is actually going to SIGN AUTOGRAPHS. He charges down the steps, pursued by the assistant den mother who is trying to heed the den mother's aforementioned rule. She will have her hands full as an aggressive-looking mob, consisting primarily of middle-aged men holding multiple balls, baseball cards, and jerseys, has already assembled.

The boy thrusts his scorecard into the pack. Tears form in his eyes as his feet are stepped on and his ribs elbowed. He is ready to give up when he feels the tug. Yes, the player has grabbed his scorecard! He lets go, the player signs and places it back into the pack.

The boy reaches for the program. Suddenly another hand appears from nowhere and grabs it. The hand belongs to a grown man with a shaved head and tattoos adorning both shoulders. One is a Confederate flag logo; the other simply reads "HOT MAMA."

Clutching the autograph in one hand, the man races up the steps and returns to his seat. Using his iPhone, he snaps a photo of the autograph and prepares to upload the image to eBay for a "Buy It Now" price of $30.

Meanwhile a security guard consoles the sobbing scout. The kid relates his story and points to the thief, who is now drinking $9 beers with his friends, showing them the autograph and freely admitting that he stole it from "some dorky-looking kid who probably came with his gay scout troop."

Three guards ascend the stairs and confront the thief, who replies with his middle finger. One guard reaches for the scorecard and the thief pushes him. The guard pushes back. A scuffle breaks out and another guard tases the perpetrator, who is rendered motionless long enough to be hauled away in handcuffs, eventually recovering enough to freely scream profanities at anyone within earshot, including the group from Shady Oaks Retirement Village. Another spectator, sitting two rows below, records the whole scene using his video camera-enabled cell phone. Within moments, he uploads the images to CNN where a graphics editor slaps on the phrase "EXCLUSIVE! BASEBRAWL!!" and forwards it to a news producer, who makes it the leading story on the Happening Now News Round Up. Half an hour later, the video has been viewed over one million times on YouTube.

Celebrity blogger Perez Hilton reports, incorrectly, that the beating "victim" is Lindsey Lohan's distant cousin. Website TMZ.com, citing "sources close to the brutal attack," reports the victim is 1/82th African American.

By the third inning, Jesse Jackson is leading a protest march outside Wrigley Field. The Rev. Al Sharpton has already appeared on every major network and cable outlet, demanding the firing of the security guards, the ballplayer, the entire Cubs front office, Bud Selig and Abner Doubleday. He also says "serious discussions" are needed within the Cub Scouts organization.

Meanwhile the scout has returned to his seat, sans autograph. Cameras hone in on him and plaster his face on television. An alert tipster calls the network with the boy's name and soon he is being identified on screen. At home, his mother wonders why the phone is ringing so much. Furthermore, she can't understand why the first call is a death threat, the second is from Anderson Cooper, and the third is from a producer for the Late Show with David Letterman.

A Cubs public relations official finds the boy in his seat. He apologizes profusely and tells the den mother that, if the child agrees not to sue, he will get a tour of the locker room and an autographed jersey signed by every Cubs player.

The scout hires an attorney.

I still have that Billy Williams autograph. I took it out the other day and saw more than a signature. I saw a world without 24-hour news cycles, without instantaneous information and without a box in every household that contains numerous adapters, hopelessly tangled like Christmas lights, which look identical yet recharge different electronic devices and will forever remain in the box because the owners have absolutely no idea what goes with what.

I could have cried. Instead, I smiled. And then I laughed. I was no longer a comedian delivering jokes about technology; I had become an audience member listening to them. As I said, I find reality funny and like it or not, technology is real. So why not laugh at it?

The following stories are not a "how to" manual for raising kids in a high-tech world. I'm not a child psychologist—heck, I'm not even Dr. Huxtable—and have no business giving advice to anyone. I'll leave that to Dr. Phil, Dr. Oz, Dr. Drew, Dr. Sanjay Gupta and

Dr. "No Real Specialty but Managed To Get Myself on Oprah." But when it comes to the serious job of communicating with your kids in a tech-laden world, technology can either be your competitor or your teammate. The only way to achieve the latter is through trial, error, more error, hair pulling, an occasional stiff drink and a heavy dose of humor. So if you find your child's 57-page phone bill slightly odd; if your four-year-old is the only person in the house who knows how to program the DVR; if you've ever tried to take a technology "shortcut" resulting in disastrous consequences; or if you're just somebody searching for values and normalcy in a world gone amok, read on.

And do something that there is no "app" for.

Laugh.

CHAPTER 2

The Joy of Sex... With 20 Specialists

Let me begin by making two factual statements:

1) This book would not have been possible without technological innovations.
2) I have excellent sperm.

Numerous women have confirmed statement number two. It's the kind of compliment you hear when you and your wife must resort to technology to actually *conceive* a child. Of course I'm talking about fertility treatments, also known as "your privacy just went out the window."

I am proud to say that there is nothing physically wrong with either Sue or myself in the baby-making department. Numerous injections, scans, X-rays and assorted other embarrassing procedures confirmed that. We were just among the millions of couples who couldn't seem to roll the correct number on the dice—no matter how hard we tried. As the years flew by, we came to the realization that pleasuring each other, while enjoyable, might never result in more seats at the dinner table. So we began investigating options. We knew plenty of couples that had undergone fertility treatments—three on our block alone. However, the old "there must be something in the water" theory did not apply because the couple at the block's end has eight kids and there are probably two more hiding in a closet that they forgot about.

Who are you?

I'm your son Alex.

Honey, when did we have Alex?

Some people can just bump into each other and produce children as easily as making toast. The night of my wedding reception, my sister- and brother-in-law decided to "try" for a third child. My niece was born nine months, thirteen minutes and forty-seven seconds later.

Natalie was conceived with help from our local pharmacist. By that, I don't mean the guy came over to our house and gave us instructions.

Greg, get your hips up a little bit. Okay, hold it right there!

Sue's gynecologist, surprised that she was not pregnant after nearly two years of "trying," prescribed Clomid, a pill so small, I figured it wouldn't even help us conceive a colony of ants, much less a child. But it obviously contained some magic because Sue swallowed a couple and Natalie was born ten months later. We thought we had struck gold. Just take this teensy pill and get a baby. We GUARANTEE it! If you're not COMPLETELY satisfied, return the unused portion in this Barbie doll purse for a FULL REFUND!

Unfortunately, when it came time to make a brother or sister for Natalie, the Clomid infomercial proved false. Sue swallowed enough Clomid to impress Michael Jackson's personal physician but came up empty. So we sought out a "fertility doctor," also known as "the guy who's mastered the art of not giggling in public." I knew I wouldn't like him because his first name was Kevin. "Kevin" doesn't sound like a doctor. "Kevin" sounds like your child's playmate or a race car driver. You don't want to be in the operating room and have somebody say, "Now it's time to make the first brain incision. Kevin, will you do the honors?" Only guys with academic names like "Robert," "James," or "Omar" should be allowed into the medical field.

After reviewing Sue's chart and determining that she was okay, Dr. Kevin wanted to make sure nothing was wrong on my end or rather, *with* my end. He handed me a cup, a list of instructions and said I could either "produce a sample" in his office or do it elsewhere and deliver it *to* his office. Kevin said this in a perfectly nonchalant manner, as if ejaculating into a jar and handing it over to strangers was as routine as filling your car with gas.

"Some people feel more comfortable producing at home," Kevin said. "That's fine, except that it has to be here in half an hour or we can't get an accurate reading."

My mind raced forward to some horrible scene: speeding toward the clinic with my most personal possessions in a plastic cup and hitting a bump. Or worse, getting pulled over.

Okay buddy, what's your hurry? And don't give me the old "I've got sperm in the car" excuse.

Nevertheless, I opted to do the deed at home, not entirely comfortable with the idea of pleasuring myself in a doctor's office where I assumed there were a bunch of female technicians directly behind the wall, having morning coffee.

When I got home, I peeked at the directions, as if the task at hand (no pun intended) needed them:

Step 1: Produce a sperm sample by the "masturbatory" method.
Step 2: Inform the clinic if there is any spillage.

Despite four years of journalism school at Northwestern, five years of reporting, two national journalism awards and numerous readings of the articles in Penthouse *Forum*, I had never encountered the word *masturbatory*. As I write this, even the spell checker on my PC is stumped. Is it an adverb? A proper noun? A dangling participle?

Secondly, I'd be damned if I was going to inform anybody of "spillage." It's embarrassing enough, I thought, to be depositing yourself into a cup without having to announce that your aim is comparable to an assassin with hiccups.

As the big day approached, I grew more and more apprehensive. Sue didn't help matters by requesting details. What room in the house did I plan to use? What time was mastubatory-ing going to take place? What was I going to think about?

The "what are you going to think about" question was the most nerve-racking. I speak for all fertility-challenged men when I say there is no right answer. Reply, "You honey, only you," and she assumes (correctly) that you're lying. Provide the truthful answer—Jessica Alba and Megan Fox sharing a twin bed in a Miami Beach hotel room with no air conditioning—and she'll burst into tears and stare daggers for a week. The only suitable answer in this situation is, "I dunno. Whatever pops into my head." Incidentally, George Bush ran our country for eight years using that same technique.

In spite of my embarrassment and insecurities, I followed the directions down to the letter. I awoke at 6:00 a.m. and tiptoed to the bathroom, careful not to wake anybody, especially the dog. That's all I needed: my Yorkshire terrier staring at me and thinking, "Hey, I didn't know humans could do that, too!"

Per instructions, I produced a sample—without spillage, thank you very much! I labeled the cup, inserted it into the Dr. Kevin-supplied brown paper lunch bag and high-tailed it to the clinic, keeping my speed at 35 miles per hour and deftly avoiding potholes. I arrived with my sack, only to see two men in front of me, carrying similar sacks. Our eyes met as if to say, "Well, I know what you were doing thirty minutes ago." I wanted to ask these guys out for coffee or a round of golf. I sympathized with them because they were wearing business suits and clearly had important places to be. My home office entitled me the option of taking the day off and wallowing in self-pity if I chose to. For these guys, producing a sperm sample was just one element of a busy day. I could envision their schedules on their Blackberries:

5:30 a.m. – *Take conference call with Europe*

6:00 a.m. – *E-mail sales report to marketing department*

6:30 a.m. – *Masturbate*

7:30 a.m. – *Staff meeting*

Once I had surrendered my sack, I was given WAY TOO MUCH INFORMATION. The sperm would be "cleaned," the bad ones discarded and the remaining ones counted. I chose not to ask questions but to silently marvel at what technology has been able to accomplish, and no I'm not talking about our ability to send Christmas cards via e-mail. Sure we can put plastic hearts in humans, restore vision with a laser beam, and even clone sheep. But *clean* a sperm? How was this possible? I imagined a lab technician peering under a microscope, holding an infinitesimally sized scrub brush and lathering the sperm, while they tried in vain to swim away. Giving a bath to a two-year-old never sounded so easy.

Discard the bad ones? What exactly separated a good sperm from a bad one? Were some deformed? Missing tails? Carrying liquor bottles and smoking cigarettes? Would their peers mock them?

Hey, did you hear about Stan? Didn't make the cut. What a wimp!

Finally, the idea of "counting" sperm intrigued me. This sounded like the worst job ever invented—worse than New York City garbage collector in August or soccer referee in Latin America. My sex education, which consisted of junior high reel-to-reel films and watching Dr. Ruth Westheimer on *Letterman*, taught me that men produce MILLIONS of sperm. Donald Trump probably produces BILLIONS! Bill Gates? BILLIONS UPON BILLIONS!! Counting them would require supreme concentration. What happens if the "counter" gets interrupted?

COUNTER: *Sixty million, five hundred forty seven thousand eight-five, sixty million, five hundred forty seven thousand eighty-six...*

OTHER COUNTER: *Hey, I'm going to the deli. Want something?*

COUNTER: *Sure, I'll have pastrami on rye and a Diet Coke. Sixty million, five hundred thirty, no wait, forty...SHIT...one, two, three, four...*

Some questions are better left unanswered. I would be content with the final number, whether it was two or two billion.

The final number was, in fact, 82 million with "80 percent motility," whatever that meant. Apparently 20 percent of my sperm were falling down drunk or inadvertently swam into industrial machinery. But 82 million? How could Sue not be pregnant when 82 million sperm were fighting over one egg? Could my sperm be that stupid? I know men are hesitant to ask for directions but this seemed ridiculous! You'd think at least one of the little guys would be armed with a road map, a GPS device, a printout from Mapquest...SOMETHING! Apparently they were not, or I would not have been handing Dr. Kevin thousands of dollars.

Now that Dr. Kevin and his staff had proven my body capable of producing another offspring—in fact, 82 million offspring on a good day—he prescribed an injection for Sue to stimulate her egg production. Then he told us to "make a date" on certain days.

Apparently, that is the medical term for "have sex." With these words, Dr. Kevin had just removed what I consider to be one of the highlights of lovemaking: spontaneity. Suddenly we were like those guys with the Blackberries and the accompanying sacks. We had to *schedule* our intimacy. It was like talking to the cable repairman.

How's Tuesday evening for you?

Uh, not good. Natalie has a swim lesson.

Tuesday afternoon?

I've got car pool duty.

Tuesday morning between 6:00 and 7:00?

Make it between 6:45 and 7:00. I'm still not awake at 6:00.

It's a date!

Not exactly something you would see on *The Bachelorette*.

Needless to say, our passion did not begin with candles, romantic music and a carefully chosen Pinot Noir. It began with one of us smashing the bedside alarm clock and groggily saying, "It's time!" It also did not produce a child. For three months we "made dates" without success. Dr. Kevin was puzzled and we were frustrated. So we did what any normal couple would do in this situation: we fired Dr. Kevin. Actually, today's normal couple would not only fire Kevin but also sue him. We elected not to clutter the legal system any further.

Instead we asked our friends for another recommendation. Since there were so many couples in our neighborhood undergoing fertility treatments, discussing it at block parties was routine. While the men talked lawn care and football, the women cheerfully compared progesterone levels.

One neighbor recommended Dr. Brian and we quickly made an appointment. Again, "Brian" is not a name that I would associate with medicine but I couldn't be choosy anymore. Dr. Brian was South African—a pleasant man with a passion for golf and a slew of pictures on his office wall of children conceived with his help. Looking at those pictures made me more nervous than excited. Fertility treatments drastically increase the chances of multiple births. Just ask Nadya "Octomom" Suleman, a single California mom who already had six kids yet took fertility drugs and on January 26, 2009, produced *eight* more. That's like Joan Rivers requesting more plastic surgery.

We knew about the chance for multiple births but the point is further driven home when you're staring at Christmas cards containing LITTERS of children. In one photo, an adorable four-year-old girl was holding her three new siblings. The look on her face seemed to say, "What's the matter? Wasn't I good enough for you?"

In another, a boy and girl, about five and three respectively, were surrounded by quadruplets. In Harry Potter novels, photographs can talk. This one would have surely said, "Guess who's not ever gonna get paid for babysitting?"

Fertility doctors must consider these photo collages to be the ultimate resume. Dr. Brian's wall screamed, "Look how good I am at what I do. You want ONE child? I'll give you FIVE! No extra charge! How's that for customer service?"

Dr. Brian suggested we try a procedure known as an "Intra-Uterine Insemination" or IUI if you're in a hurry. To put it bluntly, he would fire the sperm into Sue from a syringe, hoping to, in his words, "hit the target." That sounded like a good idea considering that 82 million sperm had been unable to find the target on their own. Perhaps a little professional nudging would do the trick. This procedure, Dr. Brian explained, would be done in his office and a fresh batch of sperm would be needed. I was handed another cup and the same set of directions. Drs. Brian and Kevin apparently shop at the same medical supply store.

Dr. Brian, we discovered, was a busy man and appointments were hard to come by. Sue secured an early morning slot in March. It was that or nothing for a month. I checked my calendar and saw a slight problem.

"You know, I'll be in Los Angeles the previous evening, auditioning for the *Tonight Show* and the *Late Show with David Letterman*," I told Sue.

"No problem. You can just fly home on the red eye and meet me at the clinic," she replied.

Of course, "meeting her at the clinic" meant I would have to produce a sample at the clinic. Or I could just take the microphone on Flight 1740 and inform all the passengers that the bathroom would be out of service for seven to thirty minutes. The latter would no doubt prompt American Airlines to rescind my frequent flyer status, so I opted for the inevitable.

My audition that evening for the two shows every stand-up comedian covets went horribly. I refuse to blame it on the fact that my mind may have been preoccupied. The audience just didn't dig me. Dejected, I flew home and instructed the cabdriver to drop me and my luggage at Dr. Brian's office, where Sue awaited. I burst in unshaven, bleary-eyed, hungry and slightly jittery thanks to a quick detour at Starbucks.

"Ready?" she asked.

"As ready as I'll ever be," I replied.

I looked around the waiting room, glad that I was the only man there. Of course, there was a small part of me that wanted to converse with another male, as I knew I would have had bragging rights.

Yes, we're both about to masturbate. But last night I met with talent scouts from the Jay Leno and David Letterman shows. What did you do? Huh?

Turns out, I was NOT the only man there. The other one was in the "production room." I would have to wait my turn. And since the production room had no time limit, I was forced to leaf through copies of *Good Housekeeping, Vogue* and *Expectant Parent* in between checking my watch. *Expectant Parent* was my favorite. There was probably six months worth of issues lying around, each containing the same articles, repackaged with new headlines. July's cover piece, "TEN WAYS TO CALM A CRYING BABY," was replaced with August's "HOW TO SOOTHE YOUR FUSSY INFANT," and September's "IT MAY NOT BE COLIC." Why, I wondered, would anybody want to have children after reading these articles? When you walk into a BMW showroom, you don't see auto accident photos, do you?

I was on my way to becoming a leading authority on crying when one of the technicians poked her head through the reception window.

"Mr. Schwem, the room is free. You can go in now," she said in a sunny voice. She handed me my now-familiar cup in full view of about twenty women who were waiting for Dr. Brian. They said nothing but I felt as if I could hear them silently having conversations with each other.

That poor guy.

Poor guy nothing. We get poked full of needles. All they get out of this is pleasure.

Yeah, who the hell does he think he is anyway?

Arm yourselves girls. Let's get him!

The nurse led me through the office, past all the employees, and down a hallway. A man about my age passed me, heading the other direction. Simple detective work led me to realize he had been the production room's previous tenant. We did not shake hands or even fist bump.

The walk was only about fifteen steps but I felt as if I was starring in the climactic scene from *Dead Man Walking*, when Susan Sarandon places her hand on Sean Penn's shoulder and says, "You are a child of God," before the lethal chemicals drip into his vein.

Did God intend children to be produced this way? Surrounded by petri dishes, test tubes and sterile plastic cups? It didn't matter. Things were clearly out of my hands (again, no pun intended) at this point.

The production room was just that—a room. I glanced at the sorry furnishings, all of which looked as if they had been through several garage sales and eBay listings. A leather recliner dwarfed the area. Why leather? Didn't these people realize that things STICK to leather?

A sorry piece of oak furniture ran along the wall, directly facing the chair. Metallic cabinets hung over a stainless steel sink. A dimly lit floor lamp illuminated the corner, in a pathetic attempt to create a mood even though there is no proper mood when making a date with oneself. A coffee table with a slight chip in it was nestled under the lamp. On the table was a single issue of *Playboy*.

One issue? I assumed it had to be the most phenomenal issue that ever came out of Hugh Hefner's mansion. I envisioned the lead story: "JESSICA ALBA AND MEGAN FOX. NO LONGER 'JUST FRIENDS.'"

The technician interrupted my thoughts. "When you're finished, write your name on the label, place the cup into the slot and push it around," she said, pointing to a revolving, lazy-Susan contraption in the wall. In a moment, my assumptions were confirmed. The sperm washers, sperm counters, sperm interns and sperm hangers-on really were DIRECTLY ON THE OTHER SIDE of the wall. There were women behind that wall who couldn't begin their jobs until I was done. Until then, they were probably throwing money into a pot.

Three to one says he takes more than fifteen minutes.

Who thinks he'll use the magazine?

How will we know?

We can hear him turning the pages. You must be new here.

I've got ten bucks that says he brought something from home.

Marcie, your bagel's ready.

Jane, the last guy took twenty-seven minutes. Didn't you guess twenty-five? Here's fifty bucks!

Wait a minute. I guessed twenty-eight.

But you went over. Remember the rules?

Ante up, girls. Round two is about to begin.

The technician closed the door, but not before placing a "DO NOT DISTURB" sign on the handle. Was that sign necessary? Wouldn't a simple closed door be a proper signal for the staff to keep out? Maybe not. Maybe the sign was added after an employee opened the door and said, "Oops. I just came in to get some supplies. Don't mind me."

I sat in the chair (yuck) and looked at the oak thing. It appeared there were doors on the front. I rose from the chair for a closer look. I opened the doors and discovered a television. Why these doors were shut in the first place, I don't know. I didn't get a chance to ask the last guy, who, chances are, was not hanging around the lobby area smoking a cigarette. Several VHS tapes lay on the top of the TV.

That's right, I said VHS tapes. By this time the DVD had replaced the VHS tape as the new mode of home entertainment. But fertility doctors apparently have more important things to spend their millions on. Brown paper bags and malpractice insurance, for instance.

I picked up several tapes and noticed that none had been completely rewound. The previous viewers had apparently used the tapes to complete their tasks, hit "stop" and fled the room in shame. I inserted one tape in the VCR and rewound about five minutes of it.

I refuse to reveal any more details. Suffice it to say that I succeeded in the task at hand (no pun…forget it). I pushed the cup through the wall and exited the production room, a sound that seemed to reverberate throughout the clinic. But first I remembered to keep the cabinet doors open so the next tenant wouldn't have to find the television on his own. Standard production room courtesy.

The IUI takes place almost immediately after the sperm receive their massage and beauty treatments. I joined my wife in a separate room, anxiously awaiting Dr. Brian's arrival. Several minutes passed before the door opened. It wasn't our doctor but another technician, armed with a clipboard and a syringe.

"The doctor will be in shortly," she said, glancing at the clipboard. "So, how are the sperm today?"

I didn't know if she was talking to the sperm, the clipboard or us. While I contemplated this, she answered her own question.

"Ninety million. Excellent," she exclaimed, glancing at me with a slightly raised eyebrow and an approving smile. Turning to Sue, she said, "He's a good one. Keep him."

I wondered what Sue's other options were. I also wondered how I managed to produce an extra eight million. Certainly it wasn't the surroundings. Maybe it was the double latte.

Alas, the eight million were just as stupid and confused as the previous 82 million. Even Dr. Brian's expert aim didn't help. The process failed.

We were down to our final option—in vitro fertilization. You know the phrase, "You can lead a horse to water but you can't make him drink?" In vitro actually forces the horse to drink. The sperm and egg "make a date" under a microscope with the help of a technician who hopefully is void of any nervous twitches. The two introduce themselves and get to know each other for three to five days in a dish. Then, if they're still hitting it off and have indeed hooked up, Dr. Brian moves the party to the uterus. Guess what? You're pregnant! The question is, will you remain pregnant?

In vitro is not only ungodly expensive but requires intense preparation. For two weeks, Sue underwent painful injections designed to increase egg production. With these drugs, women don't produce one egg a month, as is normal. They can produce upwards of twenty. The idea is to retrieve as many as possible, fertilize them all and put several back into the owner, in hopes that at least one hangs around for nine months. If they all hang around, they end up on Dr. Brian's wall.

As a man, it was impossible not to feel guilty. Why was I so uptight about my responsibilities? The women in the waiting room were right; my job was simply to provide myself with a little pleasure. Nobody was poking me in personal areas or making me lug twenty eggs around all day. I didn't go to sleep crying or wake up knowing that my day would begin and end with needles. We had heard stories of women, sky-high on hormonal enhancements, who locked their husbands out of the house, threatened them with sharp objects (including the needles) or heaved anything not nailed down in their general direction. Sue was a total trooper throughout the ordeal, never once launching any projectiles my way even though I nicked a vein one night while injecting her, producing a spray of blood that we named Old Faithful.

On "retrieval" day, Dr. Brian removed fifteen "follicles" from Sue. The next day we were told that all but one fertilized, meaning we had the potential for fourteen children and a reality show on The Learning Channel, the same network that created *Jon and Kate*

Plus Eight and its sequel *Kate Plus Eight Plus Multiple Nannies Who Watch The Kids While Mom Competes on Dancing With the Stars.* Now there was nothing to do but wait. Dr. Brian explained that he hoped the "couples" in the dish could make it to five days. The longer they survived outside, the greater the chances they could survive inside.

Indeed, two of them did make it to five days and were implanted into Sue. The waiting process began again. It would be a few weeks before Sue could take a pregnancy test. In the meantime, there were MORE injections, designed to reduce the chances of miscarriage.

In vitro patients don't pee on a stick to determine pregnancy. The process begins long before a stick would produce any valid results. Blood is drawn and estrogen levels are measured. High numbers, we were told, point to a pregnancy. Sue's numbers were indeed high when she went in for her blood draw. She phoned me in Atlanta and told me the news. We cried like the babies we hoped would soon be ours and attempted to hug each other through the phone.

Our euphoria lasted less than forty-eight hours. A second blood draw revealed alarmingly low levels, meaning Sue had suffered an early miscarriage or someone made a mistake in the lab. Our emotions were reaching roller-coaster level proportions. We had gone from the top of Mount Everest to the bottom of the Grand Canyon in two days. We were now facing the inevitable question: do we try it again? Would things be any different? In vitro doctors make no promises. You pay $10,000 and they try their best. But there are no money-back guarantees, no thirty-day trials, no "even if you're not pregnant, keep the knife set as our free gift."

After much soul-searching, we opted to give Sue's body a brief rest and then make one more attempt. Our strategy worked: Amy was born nine months later. In less than a year, we had gone from cursing technology to praising it. We had yet to be introduced to the complexities of text messaging, Facebook friends, online bullying or other problems that would soon confront us as we muddled through the world of parenting in this high-tech age. Instead, we had seen how something unheard of in our parents' generation had been perfected with the help of computers, print outs and assorted gadgets that I couldn't even begin to explain.

I love my children equally but one thing will be different. When Amy is a moody teenager, rolling her eyes and talking back at every occasion, I have the perfect comeback line:

"Remember young lady, I could have left you in the dish."

CHAPTER 3

It's a Digital World and I Just Live In It

On March 6, 1997, I acquired two things: my first child and a digital camera. Thirteen years later, I have yet to learn how either operates.

The digital camera is an integral tool for parents and, in my opinion, everything wrong with what technology hath wrought, namely, the inability to EVER be satisfied.

With a digital camera, one no longer shoots a roll of film and prays to God that at least *one* picture turns out decently. Now everyone is a professional photographer; we can shoot, delete, zoom, crop, remove red eye, stroke, purge, blur, render, sharpen, pixelate and apply every other tool created by Adobe Photoshop until we finally get that ABSOLUTELY POSITIVELY PERFECT photo of our children. This priceless keepsake is then transferred to a hard drive where it remains unviewed until the computer crashes and the photo is lost forever because we never learned how to back up data.

Up until I purchased the camera, the only digital thing I *thought* I owned was an alarm clock. But numerous websites informed me that the Digital Age actually began in the 1950s when the transistor was invented. I did have a transistor radio so I guess my entry into the Digital Age must have occurred whenever Cheech and Chong's *Sister Mary Elephant* was released as I remember listening to it over and over on my vanilla-colored AM radio.

The transistor, incidentally, also spawned creation of the words "doo hickey," "whatchamacallit," and numerous other terms that describe something we can't explain but feel we must have. I've always thought that should be the barometer for owning any piece of technology: if you need to use some generic term for it, you obviously don't understand how it operates and therefore should stay away from it. Even the companies that manufacture this stuff can't explain how their creations work yet that doesn't stop them from reaping millions of dollars while the rest of us live in a constant state of confusion. Over-the-phone computer salespeople from companies like Dell are masters at using multisyllabic words to describe things that sound cool but are only useful if you earned a Masters degree in Information Technology. From their call center cubicles in India and the Philippines they extol the virtues of "more RAM," a faster "media accelerator" and "integrated subwoofers" as dumbfounded consumers silently listen from thousands of miles away, eventually saying, "Yeah, put me down for more RAM. I GOTTA have that."

Once you own the computer, the confusion only intensifies. On a recent morning, I powered up my PC only to find the following message: STOP: 0x0000007E (0xC0000005, 0xF8390AE1, 0xF2AE3CC8, 0xF2AE39C4). It looked serious so I did the only thing I was taught to do in such dire situations: turn the PC off and turn it on again. The message disappeared.

Several years ago, I was backstage at a corporate event featuring a keynote address from Microsoft CEO Steve Ballmer. Ballmer, it should be noted, may be the world's most intimidating leader of ANYTHING. An absolute bear of a man, with a voice to match, Ballmer looks as if he could strap on shoulder pads, walk onto a football field, kick the living crap out of Peyton Manning and do it while running Microsoft Excel. If we sent Ballmer to the Middle East armed with nothing more than a bullhorn, Osama bin Laden would appear meekly from a cave and handcuff himself to the nearest tree. I know Bill Gates handpicked Ballmer to be his successor at Microsoft but I have to wonder whether fisticuffs, arm wrestling or a toilet swirlie somehow factored into that decision.

So there was Ballmer, getting miked by a scared-as-hell audio technician, looking at notes and already sweating profusely. If you don't believe the last part of that sentence, search "Steve Ballmer" using Google Images and you will quickly see that light blue button down shirts should be immediately stricken from his wardrobe.

Like most corporate speeches, Ballmer's presentation included PowerPoint slides. Perhaps he felt obligated to use them simply because Microsoft invented PowerPoint. Truth be told, PowerPoint was created so audiences divert their eyes from a speaker long enough for said speaker to pick his nose. This works by the way. I've done it.

The only thing needed to operate PowerPoint while speaking is a wireless mouse. Realizing he didn't have one, Ballmer asked a member of the backstage crew for "the clicker."

The clicker? The head of the world's largest and most successful software company was calling it a *clicker*? It was akin to Albert Einstein calling the atomic bomb "that thing that goes 'boom' when you drop it." At the very least, Ballmer should have said, "Where is my 802.11b radio frequency operating at 2.4 Gigahertz advancing device?"

Instead, it was "the clicker." Even Ballmer couldn't explain its inner workings yet he couldn't stride on stage and address fifteen hundred corporate minions without it.

I am equally clueless about most things digital. I currently own a digital TV, digital phone, and digital receiver which causes digital radio stations to blare from my outdoor digital speakers, thereby annoying the digital neighbors who are mowing their digital lawns.

I'm not sure how all this digital technology improves my life. I do know that my digital phone does not change the conversations coming from the handset, my digital radio does not make rap music any less profane, and my digital television does not make Jay Leno's monologue funnier, President Obama's Afghanistan strategy speech more intelligent, or the Chicago Bears a better football team. But "digital" is like "more RAM." It's apparently necessary for everything. Just ask the Federal Communications Commission.

Sometime in 2008, the FCC casually announced that everybody who owned a television had better GO DIGITAL by June 12, 2009, unless they wanted to stare at a blank screen every night. Consumers had two choices: purchase a television with the word "digital" somewhere in the instruction manual or acquire a DIGITAL CONVERTER BOX.

This piece of cryptic news caused thousands of senior citizens to run to Radio Shack, use hefty portions of their monthly Social Security checks on "do-it-yourself" digital conversion kits and bone up on terminology such as "set-top-box," "over-the-air" television, "backward compatibility," and "channel scan." At last count, exactly two people over sixty-five have successfully converted to digital, although both are convinced the newfangled doo hickey between the rabbit ears on their TV sets will allow them to receive *The Texaco Star Theatre with Milton Berle*.

Remember Doc, Christopher Lloyd's character in *Back to the Future*? He spent most of the movie telling Michael J. Fox's Marty that 1.21 gigawatts of plutonium was needed to power the flux capacitor. As if this wasn't bad enough, the lightning bolt had to strike the flux capacitor at exactly the right moment for the DeLorean to catapult forward to 1985. Sue and I feel the same way about our camera; if we do everything positively perfectly, we will reap the full benefits of its digital capabilities.

We are still waiting for that to happen.

Our photographic ineptitude becomes readily apparent each December when we strive to take the perfect Christmas card photo. Prior to the invention of digital cameras it was so easy to snap that holiday greeting card: take some photos, drive the film to the drugstore, wait seven to ten *business* days (apparently photos get weekends off), retrieve the prints, realize that all of them suck but it's too late to take more, choose the one that makes your children look least revolting and send out 250 copies to people you only correspond with at Christmas time.

The process becomes slightly different when you're holding a camera complete with a DIGIC III Image Processor, 12.2-megapixel CMOS Sensor, 3.0-inch LCD monitor, and compatibility with SD and SDHC memory cards!! (At least that's what the guy in the Philippines told me.) Now you can take HUNDREDS OF PHOTOS, delete half of them, view them on a computer screen that highlights your kids' every blemish, scar and abnormality, narrow it down to ten, argue so forcefully that you and your spouse stop speaking to each other, and finally realize that it is Christmas Eve and you are going to have to send the cards FedEx to ensure they will be received before the new year.

First, let me point out that our Christmas card photo never includes my wife or me. Christmas photos are supposed to be a chance for people to see how things are GROWING, i.e. your kids. They are not meant to showcase things that are RECEDING, i.e. your hairline, your abs or your boobs.

One great thing about Christmas card photos, digital or not, is that they encourage creativity. One year a neighbor put all four children in matching Christmas pajamas and posed them leaning on an oversized holiday gift box. Another couple dressed their children in Wild West outfits. As cute as these photos are, I look upon them with envy because, whether they're wearing bathing suits, evening gowns or matching Chinchilla coats, my kids never seem to photograph that well. If they looked as good as our neighbor's kids, I'd auction them off on eBay.

Personally, I think the worst Christmas photos are from childless people who, instead of sending a plain old card, send pictures of their pets. Children change; pets don't. Furthermore, children get taller. Pets die. If they are fortunate enough to live long, productive pet lives, do they look that different? Growing up, I had a cat who looked exactly the same at age two that he did at twenty-two. He'd lost all his teeth but since cats rarely smile, nobody knew.

Okay, I'll admit it, our first card as a married couple contained a photo of *ourselves*. We climbed into one of those giant ball pits featuring thousands of plastic balls that are the source of every malady from psoriasis to swine flu. The pit was located in a suburban strip mall establishment called The Discovery Zone. Basically, it was an indoor playground that charged money so kids could run around and blow off steam. My kids do that every day at home for free, thank you very much. But the Discovery Zone's founder apparently thought the play experience was enhanced if it came with a $10 price tag

So off we went one cold November day to the Zone. We told the manager our intentions, he complied and then we shooed all the kids out of the ball pit. Since Sue and I were twenty-eight and thirty-one respectively, and the average age of a Discovery Zone customer is six, we didn't get a lot of resistance from the other ball pit patrons. They stood outside the pit and sucked their thumbs while we frolicked in the germ-laden balls and giggled like the carefree newlyweds we were. The manager himself took the photos. We thanked him, left and went out for cocktails. How many people go directly from the Discovery Zone to a bar?

If you think you've taken a great Christmas card shot, send it out early. That way others can compliment you when they send cards in return. The responses to the ball pit shot were overwhelmingly positive.

You guys look sensational!

Great idea!

You two need some kids.

Like I said, the responses were *overwhelmingly* positive. I didn't say they were *completely* positive.

My parents were never much for Christmas photos. They sent them out until my older sister Julie and I were ten and eight. Then the practice subsided until my mother inexplicably revived it one year without telling us. Never one to waste film, she snapped a single shot of us posing in front of the fireplace prior to attending our high school homecoming dance (not with each other, in case you're wondering). My sister, seventeen at the time, wore a salmon-colored strapless dress which unfortunately accentuated every tan line from the previous summer.

I was fifteen, in the throes of puberty, and every photographer's worst nightmare. My tie knot was too big and my haircut too short. My suit was made of the finest polyester blend. The flashbulb didn't know whether it should reflect off my glasses or my braces so it chose both. We reluctantly posed, convinced that the resulting photo would not even make it to the side of a milk carton.

We were so wrong. I came home from school two weeks later to find three hundred copies on the dining room table, accompanied with a Santa logo and, in festive cursive, "Merry Christmas from the Schwems." A more appropriate caption? "Merry Christmas from the Dork Children!" The envelopes were already addressed, stamped and about to be sealed. A simple tongue lick was the only thing separating Julie and I from embarrassment beyond our wildest dreams.

Oh, how I longed for a match and a can of gasoline.

Unfortunately, I never found either and my mother mailed the cards the next day despite our threats to run away and become (choose one) a hooker, a drug dealer, a Hare Krishna or a Democrat.

The comments rolled in shortly thereafter.

"Love Julie's tennis tan," read one of the nicer ones.

"Greg's braces are soooooo shiny," was one of the honest ones.

"We hate you," was from the card Julie and I sent to them.

Incidents like that make you vow that, when you have kids, you'll do things differently. You will not shoot just one photo; you will let EVERYONE have an EQUAL say in the selection; you will make the Christmas card photo shoot fun, fun, FUN!

For the first five years of her life, Natalie was an only child and fully cooperative during the photo sessions. We used the digital camera but it really wasn't necessary as we always got the money shot within the first few minutes. The reason was twofold: she had nobody standing next to her to torment, and her surroundings were, to say the least, opulent as we took the photo during a week that I performed stand-up comedy on one of the approximately 5,347 cruise ships sailing the Caribbean. The earth is three-quarters covered in water; cruise ships take up more than half of that water.

That's why I'm always amazed when I read news accounts of someone who is lost at sea. Odds are excellent that, if the person simply treads water for a few minutes, he or she will be spotted by an alert daiquiri-holding conga line participant.

Incidentally, cruise ship comedian is a great gig providing you get *off the ship* at some point. I've met performers who log upwards of forty weeks a year on ships and they have lost complete touch with reality. Ask them who is the leader of the free world and the response would probably be, "The Captain."

From 1997 to 2001, every Christmas card shot included palm trees, turqoise water and one carefree daughter. Our best shot was two-year-old Natalie, in pigtails and a bathing suit, sitting at a swim up bar in Montego Bay, Jamaica and holding a virgin strawberry daiquiri. Of course our friends weren't buying the virgin part, as evidenced by the notes.

I see she takes after her Dad.

Who drove home that night?

Let me guess. She had one more and then fell off the stool.

Amy came along in 2002. I had stopped working cruise ships and therefore we abandoned the Caribbean theme. We chose our house for the next few photos, wary of bringing an infant AND her sister anywhere. We snapped the photos with a digital camera but, upon realizing that any photo session void of crying would last about thirty seconds, quickly chose our shot.

In 2005, Amy was three and, we incorrectly assumed, old enough to take direction during a photo shoot. Now the powers of the 12.2-megapixel CMOS Sensor could be used to full capacity. Sue decided we needed a real Christmasy background for this one. We dressed the kids in holiday outfits, loaded them into the minivan and drove not to a field of snow-covered pines but THE LOCAL MALL, featuring a gigantic, fake Christmas tree just yards from Santa's throne.

Note: If you do decide to use the mall as your Christmas backdrop, you should have plenty of time since most malls put up Christmas decorations around the Fourth of July. Santa appears shortly thereafter and doesn't leave until 11:45 p.m. on Christmas Eve. "Mall Santa" may be the next profession that unionizes and demands benefits and retirement packages.

The picture process is always a two-person operation. While Sue holds the camera, I stand behind her, urging the kids to "look this way" and say "poopy pickles" or some other asinine phrase that I hope will get them laughing. Instead, it merely draws gasps from shoppers and, eventually, a stern look from store security. Santa himself even pauses from his duties long enough to turn his snow white beard our way and cross me off the "good" list for the year.

We took approximately 192 shots without getting a single one suitable for a Christmas card. It didn't matter that we were using a digital camera; in 2005, my children enjoyed having their picture taken together about as much as Hillary Clinton when she is standing next to Bill. If one child smiled, the other scowled. If one laughed, the other stared at her shoes. If one summoned an adorable grin, the other looked as if she were about to get vaccinated.

Just as the head security guard was about to toss us the keys and say, "Why don't you two lock up?" we completed the task, drove home, loaded all the photos onto the PC and chose one. "Next year, they will be older and this won't be as difficult," I told Sue.

That statement proved to be about as correct as, "Mission Accomplished!" which George Bush bellowed to our troops aboard the USS *Abraham Lincoln* on May 1, 2003.

In 2006, we returned to a different mall with a different Christmas scene and an upgraded digital camera. This one could shoot continuously just by holding down the shutter, making it perfect for action shots and, we incorrectly assumed again, one stinking Christmas photo. Sue had noticed a sled, stuffed full of presents, sitting between men's shoes and women's fragrances. A perfect spot for a card, she thought.

Our kids, now four and nine, were again dressed in cute Christmas outfits and looked adorable. That is, until they climbed aboard the sled and began poking at each other, snarling at the camera and, in most "unladylike" fashion, spreading their legs just wide enough so their underwear was visible in every shot. The continuous shutter whirred in the background.

Finally, we called a brief "time out." While we regrouped and, in a moment of weakness, promised the girls ice cream if they could smile, a startling scene unfolded. A woman strode up with five children, all decked out in Christmas attire. The oldest was, perhaps, ten. The youngest looked as if it had left the maternity ward incubator two hours ago.

"Do you mind if we use the sled?" asked the mom. "I just want to take our Christmas card shot."

"Be my guest," I replied. Then I turned to Sue and whispered, "This ought to be good."

The woman whipped out a camera and a roll of film simultaneously. I watched her remove the film from its canister, unspool a slight amount, load it into the chamber and shut the camera's back door. There was nothing digital about this camera. For all I know it had been used by paparazzi to shoot Marilyn Monroe.

In the space of approximately eight seconds the kids knelt down, smiled on command and waited for the flash. Even the newborn smiled!

"Thank you," Supermom said before collecting her brood and herding them away.

Sue and I stood there with our jaws dangling on the floor. Had we actually just witnessed that? The woman snapped ONE picture. We were looking for an outlet to recharge our digital camera's batteries.

"Did you see that?" I said to Natalie and Amy. "Did you see how easy that is? Why can't you do that? Now get back on that sled, listen to Mom, smile and HAVE FUN DAMMIT!"

The kids gave me their best "my dad is a complete psycho" looks and complied. We took another two hundred shots and left. On the way home I couldn't help thinking about the mom with her antique camera and the ease in which she used it.

I vowed to never again fall for the bells and whistles that accompany a simple product. I vowed to be a bare-bones dad in a high-tech world. I would not download new versions of anything, upgrade to a faster anything or spend more money on a smaller yet more powerful anything.

Unless I need more RAM.

CHAPTER 4

$afety in Number$

At some point I realized that it was simply impossible to be a parent without upgrading from time to time. Of course getting something new means you will be giving something up in return. The ability to kiss your children, for instance.

I realized this several years ago when my wife dropped me at the airport for a trip to Dallas. Sue drove, I sat in the passenger seat and Amy, then three, sat in her technologically advanced "child safety seat," federally approved by the National Highway Traffic Safety Administration, the National Safety Belt Coalition, the National Institute of Occupational Safety and Health, the American Dairy Association, the AFL-CIO, and the Jehovah's Witnesses.

Don't get me wrong—I'm all for safety when it comes to our kids. But now it seems that everything a child comes in contact with—rattles, teething rings, little colored balls, you name it—is a potential death trap and comes with more warnings than your standard bottle of Viagra. How did I make it to my late forties? By rights, I should have been dead by age five. I spent my childhood riding in the backseat of, in chronological order, a green 1959 Nash Rambler, sky blue 1969 Mercury Meteor, mauve 1971 Ford Torino, and olive 1974 Buick station wagon. (Suffice it to say color was never a deal breaker when my parents purchased an automobile.) Not once was I bolted into a seat containing more straps than a gurney in your average maximum-security institution's death chamber. While in the car, my sister Julie and I behaved like normal children, poking and kicking each other in the backseat with arms and legs flailing as Mom deftly steered the car to and from the supermarket. We lived in Arlington Heights, Illinois, a town where there were actually other people and other cars. Any one of those cars could have plowed into us at forty miles per hour and thrown us into the next suburb. Apparently the National Highway Traffic Safety Administration didn't think this was ever going to happen, which is why we went seatless and beltless until we got our driver's licenses.

But one day our government decided children were little targets of death whenever they entered a vehicle. I'm not sure exactly when this revelation occurred but I believe it went something like this:

One afternoon a sixty-year-old congressman was driving around the Washington Beltway. In the backseat was a four-year-old child from his third marriage. The congressman was driving the child only because his undocumented Dominican nanny had the day off. While talking on his cell phone with his reelection campaign chairman, the congressman rear-ended another vehicle with his SUV. After receiving a new front fender (paid for by us, the taxpayers) and silently settling a lawsuit filed by the driver of the other vehicle for "chronic back and neck pain, long term suffering and mental anguish," our hero spearheaded a push to make children safer in vehicles, citing his own harrowing experience on the Beltway in which a vehicle, driving full-speed in reverse, "front-ended" his car. His four-year-old is lucky to be alive, he tells the news cameras, and thankfully will still be able to leave home and attend boarding school in the fall.

The congressman hastily wrote a bill, railroaded it through his subcommittee and the entire Congress, and smiled for the cameras while the President signed it into law. Because of his dedication, it is now MANDATORY that children ride in car seats until they reach age eight or become Amish, whichever comes first.

We couldn't even take our newborn daughters home from the hospital until we could *prove* that we owned a car seat. A nurse actually peered in our backseat just to see for herself. Did she think we were lying?

Quick honey, she's not looking. Throw our new child in the trunk and step on it!

The truth is, today's parents have a little more on their minds while driving than my mother did while manning her Ford Torino. Hands firmly on the wheel, her gaze straight ahead, Mom rarely blinked until the car was in park. Today *driving* the car is an afterthought. It's hard to concentrate on the road while yakking on your cell phone, flipping through a stack of Disney DVDs, getting treats from underneath the passenger seat and trying to grab a toy that has fallen between cushions of the minivan. Mandatory car seats were Congress's way of giving kids a chance to live past their first birthdays.

Knowing that we would have to pony up for a seat to avoid having Natalie, our first born, placed in an orphanage, we journeyed to our local baby equipment store two months prior to her birth. There was one several blocks away, appropriately named The Baby's Room. The store has since gone out of business for reasons I can't fathom. Babies are still around, right?

The Baby's Room was decorated like a baby's room, carried products you'd find in a baby's room—heck, it even smelled like a baby's room, albeit after the baby had been bathed and powdered. Sometimes I think those stores should smell like a baby's diaper so parents can get a whiff of what they're in for.

PARENTS: *(UPON ENTERING STORE) Ewww. What's that smell?*

SALESMAN: *It's the next three years of your lives. Can I help you?*

After checking the average price of car seats on the Internet, we made a quick stop at the bank to withdraw funds from our IRAs. Anybody who manufactures "required infant products" is certainly on the list of the Forbes 400 Richest Americans. The goal of every fledgling entrepreneur should be to either invent or market an infant product required by law. As I write this, I am sketching out plans on the computer for an ergonomically correct, American Dental Association approved, pacifier.

Once inside The Baby's Room, we were set upon by Chad, a salesman whose primary function was to make us feel as guilty as possible if we didn't select the most expensive seat. "This is your child we're talking about," was his standard line. "My child rides in this one," was his other line, as he pointed to a car seat ensconced in a glass case and protected by two heavily armed guards. Each time our eyes strayed to the more economical models, he would silently huff and roll his eyes as if to say, "Don't say I didn't warn you, you cheap, self-centered cretins."

The choices were endless. Should we purchase the "infant only" seat or spring for the "rear-facing convertible" model, knowing full well that, in two years, we would be required to upgrade to the "belt positioning booster seat." By the way, "convertible" does not mean a seat with a hood that retracts during those hot summer months so your baby can enjoy speeding down the highway, trapping bugs in its lone tooth. In baby seat speak, "convertible" means the seat can face either backwards or forwards. During an infant's first three months on earth, he or she sees only a car's upholstery while in the vehicle. Eventually the seat is turned around, causing the baby to think, "Oh, so that's how the car moves. There's a steering wheel up there!"

Whatever model we picked, I was certain that our baby would be unbelievably comfortable. The seats looked so plush; I was ready to ask Chad if they came in adult models. True, every seat's body was plastic but each was covered with soft, patterned cushions surrounded by foam padding in a variety of cool colors, any of which would be permanently stained the moment a baby threw up on them. Clip-on mobiles were available in case our baby got bored on its long journeys. The entire infant seat line doubled as "carriers," with handles that, unfortunately, required parents to bend their wrists at 90-degree angles, thereby causing Carpal Tunnel Syndrome. Chad neglected to point this out, however. He was too busy focusing our attention on other features, such as the

"Level-Right" indicator, which apparently would tell us if our baby was in danger of rolling off the seat and onto the floor.

After price comparing, arguing, crying and examining for more than an hour, we succumbed to Chad's pressure and purchased the high-end model. I think I heard alarms going off in the break room, signaling that another set of first-time parents had been successfully suckered. Then again, we got more than just a seat for our $300 (no, that's not a typo). Chad gave us a *demonstration* of the seat's features. I couldn't believe someone was explaining how to sit a child down. Should I be taking notes, I wondered? Of course not. How difficult could it be?

Plenty difficult, once Chad began his demonstration and I realized our baby would be strapped in tighter than a NASA shuttle commander.

Pay attention Mr. and Mrs. Schwem, it's actually quite simple. Just put the child in, slip this belt over its left shoulder and put this one over the right shoulder. Now lift up on this lever with one hand while pushing down the red button with your second hand. That will unlock the head restraint. Pull it over your child's head with your third hand. When you hear a "click," it's locked. You MUST hear the "click" before proceeding. Now insert these screws directly into the child's hipbones to secure the torso. Pull this sterile plastic cover over the face to keep germs out. The optional bulletproof glass should be used when riding through large, urban neighborhoods. See, there's nothing to it!

Okay, I exaggerated about the bulletproof glass, the torso screws and the plastic germ cover. But I'm sure a *Forbes* millionaire is on his way to the patent office right now with something similar. Chad gave us a doll so Sue and I could practice our newfound baby installation skills. Using our combined eight years of college experience, we tugged, grimaced and struggled with the various straps while trying to remember everything on Chad's checklist. At last we had completed the task. Or so we thought as Chad strolled over to grade us. Looking at our handiwork, he rolled his eyes again and gave the seat a gentle nudge. The doll's head pitched forward and one of its little polyurethane arms popped free of its strap. Chad looked at us and sighed. Sue and I glanced at each other and decided the money we spent on fertility treatments should have gone towards adopting a teenaged girl from an unpronounceable Eastern European province.

Eventually we learned how to operate our new seat. We even learned how to attach the accessories such as the "cup holder" and the Velcro headrest for "additional comfort while sleeping." Kids today have it made. Trust me, there was no cup holder in my mother's Torino. And a headrest? I rested my head on the window crank while trying to sleep. Today our kids know no such discomfort. Natalie is thirteen and finally free of the child safety seat but Amy sits upright, surrounded by five belts, drinking her beverage of choice and replacing it in the cup holder when she is full. However, she is unable to show her daddy any affection at the airport. When I reach inside the vehicle, she pulls her head forward in an effort to give me a kiss, only to have it snap back into the congressionally approved position. Her arms are encumbered with two safety straps so hugging is also out of the question. I could unbuckle the straps and pull her from the seat but I'm afraid I will miss my flight. Yet at least I know she is safe and if any one of ten federal agencies ever stops my SUV, I can prove it.

As a child grows, the safety issue follows parents around like the smell of bad meat, forcing them to dip further into their wallets and shell out additional money to avoid both guilt and illegal activity. Nowhere was this more evident than when we decided the Schwem backyard was ready for a swing set. After all, what backyard isn't complete without a plethora of gymnastics equipment that kills half the grass and causes Dad unbelievable levels of grief when trying to mow what's left of the lawn? Let's party!

Unlike the car seat, I owned a swing set growing up. Of course it bore zero resemblance to the technologically advanced swing sets of today. Mine consisted of multi-colored aluminum poles, soldered together by one of my Dad's handyman buddies who assembled it in return for a twelve-pack of Schlitz. Two swings that looked like oversized picnic baskets hung from those sorry-looking poles. Smaller poles held up a two-person "glider", which was nothing more than a park bench that had been sawed in half. Two kids sat on the benches, facing each other. Both then leaned back, one at a time, setting the glider in motion. Eventually it swung high enough to cause one occupant to pitch forward into the other's lap. With two riders tangled on one bench and none on the other, basic laws of physics (or was it geometry? I always got the two confused) caused the glider to slow down and eventually stop.

That was it. Two swings, a glider and a slide—actually a strip of sheet metal that dented easier than an empty beer can and heated up to approximately 500 degrees once the calendar reached June. All of those slides have probably been confiscated and melted down for dental fillings by now. But the slide—and the accompanying rickety swing set—was all we had and it lasted twelve years. We swung on it, hung upside down on it, used it as home base for games of tag and posed for photos on it. Today my parents would probably be arrested and charged with attempted murder simply for owning it. They would shuffle toward the judge in leg manacles while he sternly read off the charges confronting them:

No recessed bolts.

No rounded off corners.

No vinyl coated safety chains.

Plead guilty now Mr. and Mrs. Schwem and save yourself the embarrassment of a costly, time-consuming trial.

Phrases such as "recessed bolts" are what you hear today upon entering the "swing set showroom" which is where little swing sets are born. Then they grow up to be big swing sets that come with finance plans.

We made our first showroom visit when Natalie was three and Amy not yet born. We loaded Natalie into her FEDERALLY APPROVED, BELT-POSITIONING BOOSTER SEAT and drove fifteen minutes north to an industrial office park. I was sure we were lost. Why would a swing set supplier be located here? I envisioned an open field in the middle of a country town, where swing sets would be on display outdoors and we would cheerfully pick one with the same festive spirit that came each December when choosing a Christmas tree. A man named "Clem" would handle the transaction, happily shaving off two hundred dollars just because Natalie looked like his now grown daughter. All three of us would tear up simultaneously.

At least that's how I envisioned it.

The reason swing set showrooms are now in office parks is that swing sets have become a BUSINESS. Therefore, the swing set store needs to be in a business environment. Upon finding the entrance, we walked in and watched our daughter's eyes grow as big as UFOs. At least seven swing sets—or "clubhouses" as the company referred to them—were assembled on the same type of plastic grass commonly found at driving ranges. About ten children, desperately in need of Ritalin, flung themselves from one clubhouse to another. A birthday party was actually taking place in the showroom! And all this time, several salespeople were standing by, holding clipboards and drooling visibly, while waiting to pounce on unsuspecting prey such as us.

A middle-aged, matronly woman named Denise quickly introduced herself as our "personal assistant." How could she help us today? I replied that we were looking to buy a swing set. Denise gave us the same look that a Starbucks employee produces when you say you want a cup of "coffee." Clearly we had not done our homework on this issue and Denise began peppering us with questions. How big was our yard? How old was our daughter? How many children did we currently have? How many did we expect to have? How many neighbors' children were we planning to invite over? How much did Sue and I weigh? (I'll explain that one later.) And the kicker: How much did we want to spend?

At today's swing set showrooms, a thousand bucks will probably get you an old tire. One model, that Denise conveniently leaned on while talking to us, was *on sale* for $25,000. It was aptly named the King Kong Clubhouse because it was at least as big as the ape and the Empire State Building put together. I could send

Natalie out to play on this thing and not find her for years. However, she'd survive just fine because the swing set included a full kitchen, plumbing facilities and a high-speed Internet connection. If you want to see something similar to the King Kong Clubhouse, look no further than the White House lawn. President Obama's kids play on a swing set that looks as if it can hold five kids, four pets, three Secret Service agents and one summit meeting.

I glanced over at our daughter, who was plummeting headfirst down a $1,700 slide and decided that I would go no higher than $2,500. That would leave a little left over in the Schwem budget for other luxuries such as food and electricity.

With this figure in her head and clipboard in hand, Denise led us to the corner of the showroom where the lonely "Sunshine Clubhouse" awaited our inspection. No children were playing on it even though it was larger than my first apartment. So we listened intently while Denise ran off a checklist of facts regarding the safety of this particular set:

It's made of quality-milled redwood. (What, no pipes?)

It contains vinyl-coated steel and "technologically-advanced" plastic.

It has round ladder rungs for more "structural integrity."

It features the strongest swing clips on the market, containing a "bronze bushing" that greatly reduces noise and friction.

It can hold up to four hundred pounds.

The last fact puzzled me. Four hundred pounds? How many children did Denise think we were planning to have and what did she think we planned to feed them? I made a joke to Denise regarding the weight and she looked at me as if I were going through airport security and casually remarked that I had a bomb. Clearly, clubhouse shopping was not a joking matter. Denise explained that this particular clubhouse was designed to hold the weight of two adults as well as several children. That way, if Natalie ever got scared while playing, both Sue and I could rescue her at the same time. One thing Denise was not short on was potentially tragic hypothetical situations. I also started to suspect she was married to a guy named Chad.

Once Denise was convinced she had successfully beaten the safety issues into our heads, she turned our attention to the basic features we were staring at. The model on display featured two SAFE swings. An EQUALLY SAFE rope swing, consisting of a bright yellow rope tied to a red oval disk, hung from one end. At the other end, a six-step ladder jutted out. Denise explained that the ladder was precisely angled at 60 degrees. Should Natalie slip, Denise said, this angled effect would "catch and reposition her."

Now I've slipped off many a ladder at my house while trying to do everything from cleaning gutters to hanging Christmas lights. Not once have I ever been caught and repositioned. Instead, I end up in a heap on the cement, spew a seemingly never-ending stream of profanities, and continue my work. I neglected to mention this to Denise, who was clearly on a roll.

The precisely angled ladder led up to a "play deck," surrounded by so many "vertical safety rails" that it gave the appearance of a jail cell. A canvas roof, also known as the "penthouse," snapped onto two wooden beams. I watched as Natalie scampered up the ladder and sat on the play deck. At least I think that's where she was. Together, the roof and safety rails formed an impenetrable fortress so tight that I couldn't even see her. I'm sure she was up there thinking, "Wow, when I get older and start sneaking liquor out of the house, this is where I'm going!"

Continuing the tour, an UNBELIEVABLY SAFE "scoop slide" poked out from the play deck's other side. The "scoop" label apparently referred to the slide's shape, which looked like a loosely rolled-up newspaper. The sidewalls were high for, you guessed it, ADDED SAFETY.

The price for this set was $2475. How nice, I thought, to actually come out twenty-five bucks ahead. Of course, that was before Denise began mentioning installation charges. It would cost $450 to have two clubhouse "professionals" come to our yard and put this thing together. Or, Denise said, we could do it ourselves. I envisioned a truck dumping enough lumber to fill up your average Oregon sawmill onto my driveway along with some screws and a five-hundred-page set of directions written in English, Spanish, French, Japanese and ancient Egyptian hieroglyphics. Tack on another $450.

I was about to pull out my Visa card and complete the transaction when Denise produced the accessories page. Apparently we were looking at the "bare bones" Sunshine Clubhouse. It was like purchasing a car with an AM radio. This company had made sure that each clubhouse came with more options than a Stealth fighter. The Sunshine Clubhouse was designed to "grow" with your child, Denise said. That was a nice way of saying Natalie would probably be bored with the set after fifteen minutes and would plead for extra pieces. I looked over at her again as she navigated a $1,250 "rock-climbing wall." While I nixed that on sight, there were at least twenty other accessories to consider. Should we spring for a crow's nest? A bubble panel? A chin-up bar? A tic-tac-toe board? How about a crawl tunnel? My mind was racing and my wallet was sweating. Denise had just reeled off an extra grand in options. I wanted my old swing set with the multi-colored pipes back. You couldn't add any options to it. Try and attach a crawl tunnel and the whole thing would have collapsed faster than Lehman Brothers. By now Denise had moved in for the kill and I had to put a stop to it.

In the end, we opted for three accessories: a pair of plastic binoculars that magnified nothing, a ship's wheel that turned but moved nothing, and a set of rings that Natalie could hang from upside down, causing her to regurgitate lunch. I was clearly annoyed that we had gone over budget. Denise should have sensed this. But she reared back and fired one more pitch toward the plate.

"Have you considered installing a base under the system? We recommend wood mulch, rubber mulch or pea gravel. It makes for a safer play area."

By now, Denise should have feared for her OWN safety. At this point I didn't care if the Sunshine Clubhouse was built over a pit of molten lava or live alligators. I wanted out of there.

One week later, a large truck backed into our driveway. Two rough-looking men emerged and began hauling approximately 27,000 pieces of quality-milled redwood to our backyard. In three hours—the same amount of time it takes me to hang a picture frame—the Sunshine Clubhouse was fully assembled and neighborhood kids we'd never even met descended on our yard like flies at a union picnic. In the middle of the pack was Natalie, bounding from one accessory to the next, delighted to be the center of attention. Sue and I sat on the patio, holding hands and realizing that we had just completed another chapter in parenthood. Our wallets were lighter but our hearts were fuller. Above all, thanks to technology that once again was thrust upon us without our consent, our child was safe.

This heartwarming *Extreme Home Makeover* episode lasted about ten minutes. Suddenly Natalie broke free from the pack and came running up to us.

"Mommy, Daddy, can we get a trampoline?"

FRUSTRATIONS

CHAPTER 5

Hey, I'm Tiger Woods! Wiiiiiiiii!!

Parenthood is a series of depressing realizations. For example, realizing that even the hippest, most expensive designer fashions won't hide your increasing midsection any more than sale items from Wal-Mart; realizing that, even if you cure cancer and heart disease in one afternoon, your children will still think you're a moron; and realizing that those same children will discover new technological innovations long before you. The only way to deal with the latter is to use what I call "The Rosetta Stone Method."

Rosetta Stone is the foreign language program that, according to its website, teaches you to speak the foreign language of your choice "through a fully immersive environment right on your computer." Go to the website and you will be entertained by videos such as "Mark and Jen learn Russian," featuring a thirty-something couple lying face down on a bed, laptop between them. We hear an offscreen voice say, "MOO-she-na," whereby Mark repeats, "MUH-she-na" as Jen smiles in agreement. We never find out what "MOO-she-na" means but Mark and Jen appear positively giddy, convinced that, by the time their plane touches down in Moscow, they will be fluent enough to join a Russian protest march if they so choose.

I've never tried Rosetta Stone—only seen its kiosks in airports —but I have my doubts. The course seems like a perfect solution for impatient Americans who don't want to take the time to learn a foreign language. We'd rather *skim* one, convincing ourselves that we can function perfectly for a month in Germany armed only with the phrase *zwei Bier bitte* (two beers please). Actually, that worked pretty well for me when I visited Germany after college—until the fourth day when I was too drunk to form sentences.

Once, while perusing the website of an association involved in road and bridge building, I happened on a pamphlet, *Perfect Phrases in Spanish for Construction*. Basically, it was full of phrases that a non-Spanish speaking foreman might say to his Spanish-speaking crew. Now, a construction site seems like a fairly intense and very dangerous work environment. I highly doubt that a foreman has time to flip through a book, hoping to find the appropriate syntax for the phrase, "WATCH OUT FOR THAT CRANE!"

Nevertheless, I used the Rosetta Stone method when the Nintendo Wii hit the market—meaning I gave myself a crash course in the game, hoping I could learn enough about it to decide if it was appropriate for my kids.

Even though I am approaching the mid-century mark of my life, I still grew up in a video-game-laden world. When I was eleven my parents relented and purchased the hottest game in America at the time. This was a huge step for them as they had never been big believers in following the masses or, unlike me, immersing themselves in whatever happened to be the latest and greatest anything.

While other kids sported the coolest jeans, I was wearing Sears Tuff Skins. When it came time to choose a bike, the ten-speed models were off limits, for reasons never explained to me. Instead I pedaled furiously on a turquoise-colored Schwinn with a seat shaped like an enormous banana. My new mode of transportation had a single speed, one that Schwinn engineers deemed suitable for going up AND down hills. If Lance Armstrong were to ride that bike, he probably would dismount after five minutes and say, "I've had chemotherapy and I've ridden Greg's bike. Chemo is *slightly* worse."

Which is why it was a great surprise when I discovered, under our Christmas tree in 1973, a box labeled "PONG."

Pong represented America's foray into the world of video gaming once it was determined that kids would go absolutely bonkers watching two white sticks on a TV screen with a small white dot bouncing between them. The sticks were supposed to be rackets and the dot was, if you used every inch of your imagination, a ball. By twisting knobs on a control console, the players moved the sticks up and down, hoping they would collide with the dot and send it ricocheting to the other side. When the "ball" got past one of the "rackets," the player scored a point. An annoying "pong" sound accompanied each collision of dot on stick.

My sister and I played *Pong* until our vision was so blurred that everything resembled a white dot. I once spent three hours playing and then walked away from the game, only to feel as if an army of snowballs was headed my way even though it was August. Yes, *Pong* was truly a hit.

Realizing that American kids were very content to sit inside staring at television screens all day, game developers got to work creating new and more addictive games. *Pong* was followed by *Space Invaders*, which gave way to *Pac-Man* and *Donkey Kong*. These games were cute and served their purpose of giving kids wrist and elbow afflictions by the time they hit puberty.

But the gaming industry wanted more. Yes, it was time to actually *experience* the game rather than simply play it.

Enter the Nintendo Wii.

Launched in 2006, the Wii is to the current generation of kids what *Pong* was to '70s children. The only real difference is that *Pong* was played for hours while today's kids play Wii for weeks, sometimes without stopping to eat, use the bathroom or get on the school bus.

Because Wii was the hottest, most popular, most HAVE TO HAVE IT OR I'LL DIE game, it naturally was item number one on both my kids' Christmas lists. Santa delivered it on December 25, 2007, and I didn't see my kids again until New Year's Eve. Privately I wanted them to continue playing until the ball actually dropped in Times Square just so I could say, "Hey you two, I haven't seen you in a YEAR!"

Ha ha, that dad guy. He is hilarious!

I have read several articles stating that Wii promotes hand-eye coordination and even passes for exercise among today's young couch potato set, simply because you can stand up to play it. Much like the *National Enquirer* quotes "close personal sources," these articles attribute the exercise theory to "video gaming experts." Now, there's a job I would love to have! Still, getting kids to stand today *is* a major accomplishment. Now if we could only get them to look directly at us when engaged in conversation.

Not only did my kids stand but they quickly improved their elbow muscles by swinging the Wii remotes around like maniacs. In no time they were experts at every game on the *Wii Play* disc that Santa also dropped off. I watched with a mixture of amazement and confusion, wondering if I was doing the right thing by letting them indulge in Wii versions of pool, skeet shooting and ping-pong which, by the way, looked a lot cooler than '70s *Pong*.

Should I join them?

Only if they asked.

They never did.

Eventually the kids began stockpiling Wii games with the same speed they collect hair scrunchies. I can't walk five feet in my house without seeing a colored, twisted, discarded ponytail holder on the floor, the steps, or in unexplained places like the bushes outside my front door. Seriously, if I see one more scrunchie in the bushes, I'm going to build a vanity on our front lawn.

Wii Play gave way to games with more outrageous instructions and goals. Their current favorite is *Mario Kart* which, as far as I can tell, is something every parent should destroy once their child begins driver's ed. Players race through different tracks, employing tactics such as "Mega Mushroom," which allows them to flatten opponents, and "POW BLOCK," which causes opponents to spin wildly out of control.

Compared to *Mario Kart*, NASCAR is like idling in neutral.

A year went by; my kids eventually came upstairs for brief intervals and I became convinced that Wii would not stunt their growth or cause the entire family to appear on *Dr. Phil.* Intrigued by their addiction, I wanted to play Wii but avoided it for the same reason I passed on cocaine in college.

I was afraid I might actually *like* it.

When *Pac-Man* and its daughter, *Ms. Pac-Man*, hit the arcades in the early '80s, I pumped a year's worth of tuition into the coin slot. I just couldn't get enough of that large yellow-mouthed dot racing through a maze eating smaller yellow dots and, occasionally, bananas and oranges. *Pac-Man* was my addiction and I kicked the habit only after graduating, getting a job at the *Palm Beach Post* and coming to the realization that the newsroom would never contain a *Pac-Man* machine. That would have made for some interesting conversations with my editor.

Hey Schwem, the cops just found a body in a forest preserve. Get moving.

Hang on, I'm on level nine and I'm bearing down on the watermelon!

Now I work from home, my office being exactly one level above the Wii console. Often as I attempt to mine humor from a blank PC screen, I hear the Wii beneath the floor, making mind-numbing Wii sounds. Yet my annoyance usually gives way to a smile as I hear my daughters shrieking with delight and playing happily together, without the arguments that invariably accompany sibling activities.

Finally the temptation became too great. I decided to investigate the Wii phenomenon further, productivity and potential addiction be damned!

One evening, while accompanying my daughters to rent a movie at a local video store—an event that ALWAYS results in an argument—I passed a lengthy row of Wii games that could be rented for the steep price of $8.99. On the lower shelf, smiling up at me with a club in his hands, was none other than Tiger Woods, professional golfer and star of *Tiger Woods PGA Tour 09*.

It should be noted that this game debuted *before* allegations surfaced that Tiger Woods had banged every woman this side of Wilt Chamberlain. Although Electronic Arts, the game's manufacturer, elected to continue its partnership with Woods, it is considering packaging future editions of *Tiger Woods PGA Tour* in brown paper bags.

I bent down to examine this game that, according to the liner notes, promised the feeling of playing golf like Tiger Woods. After glancing around the store to make sure none of my neighbors had entered, I snatched the game and made my way to the checkout counter, determined to improve my game with Tiger's help.

Now I have never been a huge fan of Tiger Woods, and that was before his sordid off-course activities surfaced. Maybe it's because he makes a frustrating game look so easy; maybe it's because he married a supermodel; or maybe it's just because he's named "Tiger." Woods's dad Earl definitely knew what he was doing when he nicknamed his son. No other ferocious jungle animal sounds as cool when applied to a human. "Lion" Mickelson and "Jaguar" Els just don't have the same ring as "Tiger" Woods.

Even Woods himself has abandoned his real name, Eldrick, as evidenced by the voice mail he allegedly sent to one of his extramarital conquests—a message that began with, "Hey, it's Tiger."

If he'd only said, "Hey, it's Eldrick," the tabloids might have left him alone. Even better, he should have said, "Hey, it's Phil Mickelson!"

On a muggy summer morning, while alone in my basement, I entered the Wii revolution, hoping to play eighteen holes at Pebble Beach courtesy of the Wii and under Eldrick's watchful tutelage.

I was fortunate to play the real Pebble Beach in 2005, presenting a round to my dad as a Father's Day gift. Getting a tee time meant making numerous phone calls and finally securing a slot *four months* in advance. It also meant forking over $850 for five hours of enjoyment (and pain, considering I decided to play some of my worst golf that day).

Ten seconds after launching *Tiger Woods PGA Tour 09*, I realized this was going to be more difficult, more time consuming and quite possibly more expensive than playing the actual course. I say more expensive because I figured I would have to rent the game at least eighty times before mastering its intricacies.

First I was required to stare at the screen and learn WHAT'S NEW. In other words, what did the '09 version have that the '08 version didn't have other than a higher price? Of the many features, the most intriguing to me was thc addition of Hank Haney, Tiger's real life coach at the time. The on-screen graphics promised that Haney "will help improve your skills with the new Club Tuner features which will allow you to fine-tune your swing with each type of club."

Keep in mind that this is Wii golf I was about to play, as opposed to actual golf.

Nevertheless, a free lesson from Haney seemed like a bargain for $8.99. I had just played an actual round of golf the day before. Although I hit the ball decently, my putting was horrible and my short irons weren't that accurate. Perhaps Haney could correct these flaws.

Upon entering "tutorial mode," Haney told me I needed to "master the control." By that, he meant I needed to learn how to take a full swing. This seemed impossible considering I was "swinging" the Wii remote, a skinny piece of white plastic slightly larger than a candy bar. Swinging a Wii remote and convincing yourself it's a golf club is like eating at Long John Silver's and pretending you're having lobster.

Haney firmly told me I would only complete the lesson if I hit the ball 280 yards into the fairway—with a three wood.

I have never done this in my life, even when playing high-altitude courses and landing tee shots on cart paths heading downhill.

My first shot traveled 215 yards into a ravine, proving that Wii was nothing if not realistic. The Wii pop-up text read, "Good try but you can do better." Undaunted, I split the fairway with my next shot, which traveled exactly 280 yards. I was done with lesson one.

Lesson two from Haney covered (are you ready?) actually *aiming* the ball. I would have thought this might have been covered in lesson one but Haney had other ideas. This time the Wii instructed me to "zoom and press the A button. Tap the A button twice to jump directly to the target circle. Move the target circle by holding the B button and dragging it to the desired location. You may also aim by pressing the Control pad up, down, left or right."

Up until now, my only golf instruction, courtesy of my Dad, had been, "head down, knees bent."

As I was reading the on-screen text, a silhouetted figure of Woods himself hovered behind it. I saw Woods address the ball and then step away as if he were about to scream profanities at a photographer or overzealous fan. Again, Nintendo succeeded at making Wii Tiger as life-like as real Tiger, who has been known to emit "F" bombs after poor shots and berate fans that dare to photograph the world's most famous athlete as he works.

Then again, maybe he stepped away because he spied a hot babe in the crowd.

The text continued: "Press the A button to zoom to the target circle. Move the target circle to the green and aim for the flag."

I swung and hit the ball straight but over the flag, over the green and probably over the parking lot behind the green.

"Try again," the Wii said.

My next dozen shots were carbon copies. And here is where I realized how annoying this Haney guy was. I couldn't move to the next lesson until I had mastered a straight, accurate lob wedge shot to the green. In vain I looked for the "skip this section," "advance," "forward," or "kill Haney" buttons but to no avail. I pushed A. I pushed B. I pushed A and B together. Nothing. My only options were to keep trying or rip the Wii cord out of the wall.

I concentrated with intensity that would have made Woods and Haney proud. I also pressed the B button and somehow dragged the target circle. This proved to be excellent strategy as my thirteenth shot hit the green.

"Great shot," said Haney.

Lesson three involved hitting a manual draw/fade. In golfer's terms, a draw makes the ball go slightly left while a fade sends it slightly right. Notice the term *slightly*. Unfortunately, your average golfer doesn't do anything *slightly*; he hits the ball far left, a hook, or sends it careening wildly right, also known as a slice. But Haney seemed to think I could make my ball dance with the grace of a Michael Jackson moonwalk if I just followed the simple instructions:

"Press the A button to zoom to the target circle. Point to a colored Draw or Fade handle and hold the B button. Move the target circle left or right for desired draw or fade."

My target to complete the lesson was the fairway beyond the trees. In other words, I was supposed to *draw* (hook) the ball around the trees and land it in the fairway.

I swung the Wii candy bar and watched in horror as my ball passed through one, then two, then three patches of trees. An all-too-familiar rustling sound accompanied each collision.

"Try again," Wii stated.

I dunked two balls in the Wii creek before landing in the elusive fairway. Only then was I told that I could avoid the entire "press A, hold B, drag circle" approach and draw and fade the ball simply by twisting the Wii remote to the right or left after I swung.

I glanced at my watch. I had been under Haney's expert care for slightly more than an hour. The lesson was far from over.

In the next few minutes I learned to put spin on the ball for more distance. I learned how to hit a shot with partial power, which basically meant not swinging the Wii with all my might.

Finally I moved to the final lesson…putting. I knew this was going to be the most complex Wii movement because there were six tutorial screens. My TV showed a putting green. A grid draped the putting surface and blue, yellow, green and red dots danced across it, moving at different speeds and rolling in different directions. It looked like I was attending a strategy session at the Pentagon.

Alas, I was just learning to putt.

Haney translated the grid: Green meant the putting surface was flat. Blue meant a downhill slope. Yellow an uphill slope and red an EXTREMELY uphill slope.

I could "preview" the putt by pressing the "minus" button on the remote or simply clicking the on-screen "Putt Preview" button but was sternly informed that I'd get only one preview per putt. If I requested another preview, I assumed Tiger would appear and swear at me.

Haney's instruction continued: "Point the Wii remote down and follow the *natural* putting motion. (Haney had obviously never seen me putt.) Using the recommended percentage on the HUD, try to match that to the percentage on the Putt Power Meter."

I searched the screen in vain for the meaning of "HUD" but never found it. I assumed it was the thermometer-shaped thing that hovered on the left side of the screen. A bar rose and fell depending on how far back I drew the remote. Haney instructed me to "make the putt and begin your golfing career." I was overjoyed knowing that my lesson with this golf tyrant was about to come to an end.

And it did end, but not before I hit a five-foot put 3.2, 2.9, 1.4 and 4.7 feet before holing it on the next try. At long last I was ready to play Pebble Beach.

I selected "one golfer" mode seeing that there was nobody else in the basement. I also chose "stroke play" since I had no idea what "Rings" or "Stableford" meant. I just know I have never turned to my golf buddies on the first tee and said, "Guys, let's shake it up and play Stableford today."

I also could choose from a litany of pro golfers with names I recognized: DiMarco, Furyk, Goosen, Parnevik or the great Woods himself.

I opted for the world's most entertaining golfer, John Daly, he of the massive drives, numerous divorces, multiple addictions and skyrocketing weight. I couldn't wait to see how lifelike Nintendo made him.

Nintendo, it turned out, was very kind. A slimmer-than-I-have-ever-seen Daly strolled to the first tee, without either a cigarette or a Budweiser. As he did, the Wii announcers made their audio debuts.

"Good afternoon. Kelly Tilghman here for EA Sports," came the voice of a woman I assumed was Kelly Tilghman. Moments later she was joined by a Scottish chap named Sam Torrance.

I would soon grow to despise these two more than Haney.

My loathing of them did not start immediately as I (or should I say Daly?) addressed my first tee shot and swung mightily. Even Torrance was impressed.

"This should work out good. Down the right side of the fairway."

The Wii crowd roared. Actually roared! Haney's lessons were paying off.

I had 110 yards to the pin. While I pondered my options, a "caddy tip" popped on-screen:

Use your spin to help the ball towards the target on the green. While your ball is in flight, choose a direction on the control panel and shake the Wii Remote to generate spin.

I swung. I shook. I hit the ball 120 yards; just on the green's back edge. Daly looked dejected but again Wii was kind, as the real Daly would have dropped his club, withdrawn from the tournament and hitchhiked to the nearest Hooters.

Now it was time to chip. The bizarre looking series of balls slithered across the green, moving at alternate speeds and direction. I studied them intently, for reasons I could not explain.

"Please just swing," said Sam.

No kidding. Apparently I was *irritating* Mr. Torrance. Flustered, I swung, rolling to about six feet. Time to putt.

I gauged the putt, checked the still unknown HUD thingy and stroked.

"That's going nowhere," said Tilghman. "This is for bogey."

I putted again, moving the ball approximately eight inches.

"I've seen many poor putts in my day and this ranks right up there." Tilghman droned.

I vowed then and there to someday track down Kelly Tilghman and ask why, if she was such an expert, wasn't she playing on the LPGA tour as opposed to whoring herself by doing stupid Wii commentary?

Torrance will also become a stalking target, a decision I made once he said, "He's just trying to get out of here with a double."

I putted three more times, never moving the ball more than one foot. Even Tilghman appeared to turn sympathetic.

"This to finally end a terrible hole," she whispered as I stood over my eighth putt. That one failed as well and I was told I had exceeded my shot limit. The Wii scorecard showed 11 as Daly trudged to the second hole.

I decided to end my round there. I'm not sure if it was because of the 11, the barrage of insults from Tilghman and Torrance or the fact that I had spent just under two hours playing one hole.

I came away with two pieces of information: first, a better sense of what exactly the Wii is. I am fairly certain it will not endanger my children as long as they don't take advantage of the Internet connectivity function, which could allow them to play *Mario Kart* with unseen Wii perverts in cyberspace.

Secondly, I now possess the knowledge to submit my own game idea to Nintendo. Once it's approved and available at the video store, I will invite my kids to join me but I doubt they will have any interest in playing... *WEEKEND HACKER GOLF 2010*.

Instead, I will invite three friends over to give them the experience of playing *real* golf. We will meet in my basement some time in February, when actual courses in my town are buried under mountains of snow and changing a furnace filter counts as a recreational activity.

From the moment I fire up *Weekend Hacker*, players will see obvious differences. For starters, there will be no tutorial from Haney or any golf professional for that matter. The reason?

Weekend hackers don't take lessons, have never taken lessons and will never take a lesson until the day they die. Lessons cost money and that means less bling for beer.

Instead, we move right to choosing a course. In fact, we can choose from a litany of *public* golf courses. We scroll through the list, eventually agreeing on "Wounded Eagle." The on-screen graphics give us a tour of the first hole, a 380-yard par-four featuring burned out fairways the color of straw, a creek with no water and a green littered with ball marks, dying grass and several randomly placed cigarette butts.

We choose our players: I opt for "Sal," a twenty-year plumbing veteran whose golf wardrobe consists of a yellow tank top, frayed jean shorts and golf spikes worn without socks. My next-door neighbor chooses "Frank," a retired police officer who has played every day since leaving the force yet still can't break 100. Perhaps it's the unlit cigar dangling from his mouth that is interfering with his swing.

My golfing buddy from across town selects "Shanks." His bag consists of two drivers, thirteen irons, four putters and a hybrid 6-wood that he recently purchased from Craigslist. Shanks also plays exclusively with outrageously expensive balls designed to "fly longer, truer and straighter" than other balls. Unfortunately they still sink in water.

Finally, my neighbor down the block prefers "Wes the Press." His pockets contain tees, coins, ball marks and a roll of hundred dollar bills. Game on!

Our foursome ambles to the first tee. Before anybody takes a swing, the "gambling assistant" window pops on-screen. "Do you want to play a five-dollar Nassau with a press? Click A for yes."

Wes clicks A. Another screen appears.

"Do you want to play sandies, barkies, greenies and Arnies?"

Shanks clicks A. Game officially on!

Sal steps to the first tee, addressing the ball with utmost concentration. As he takes the club back, Frank's cell phone erupts. Sal's shot is a towering slice that hits three trees and comes to rest near the refinery that borders Wounded Eagle's first fairway.

Frank is next. Chomping down tightly on his cigar, he stares at the ball on the tee for what seems like an eternity.

"Today," Wes yells.

Frank swings. The Weekend Hacker distance meter measures his shot at forty-five yards with a tailwind.

Shanks chooses his new Taylor Made R9 driver and unleashes a blast that appears straight yet takes a hard left ninety yards in flight and ends up in a church parking lot where Mass is just letting out.

Luckily one new feature of *Weekend Hacker 2010* is the MOT, short for "Mulligan Off Tee." Each player can use it once. Shanks selects MOT and uncorks precisely the same shot, nailing a very lifelike Father O'Reilly square in the forehead.

Finally it is Wes's turn. His drive is pure and straight yet bounces off an exposed sprinkler head and caroms into the trees.

As our foursome scatters to play our second (or in Shanks's case, his third) shot, another new feature appears. It's JESSICA THE CART BABE! Appearing out of nowhere in a golf cart, and wearing short shorts and a white halter top sans bra, she dispenses Budweisers, Bloody Marys, sunflower seeds and cigars for our group and promises to return by the third hole.

The "choose a politically incorrect name for Jessica" screen appears. We can call her "sweetheart," "honey," "little chicky," or "talent" for the rest of the round.

Eventually we make our way to the green. As Frank stands over a twelve-foot putt, the gambling assistant appears again. "Would anybody like to press?"

Wes clicks A.

Frank puts down his cigar, and slightly rearranges his line, aiming for the spike mark six inches outside the hole. He strokes the putt and nails it! His triple bogey seven wins the hole. Sal and Wes had snowmans and Shanks "picked up" on the advice of the "slow play wizard," another new feature.

The second hole is the number one handicap hole, due in part to the construction crane from the soon-to-be-completed condominium development that runs along the fairway on the right. The crane comes into play on a slice.

Our foursome has plenty of time to ponder their options. For the slow play wizard pops on-screen again:

"There is a forty-five-minute backup on the second tee. Please be patient."

We search in vain for Jessica. She has not returned.

She is texting Tiger Woods.

CHAPTER 6

Santa Doesn't Bring Cell Phones... Unless They're 50% Off

Christmas is always a great time to put the Schwem "know about the technology you are entrusting to your kids" theory of parenting into effect. Just as the Wii topped their list in 2007, the following year's lists made Sue and I realize our children have one thing in common with the U.S. Department of Defense.

Both want high-tech items that will eat up an entire budget.

Okay, Natalie and Amy have yet to request unmanned drone aircrafts (although I'd like one for spying on the neighbors) but it's all relevant. If my children requested a simple board game like Scrabble, it would be tantamount to the Secretary of Defense telling Congress our soldiers could win the war in Afghanistan if they just had better thermal underwear.

There are two ways to go about purchasing your child's dream gift if said gift falls into the "electronics" category, meaning it contains numerous adapters, batteries, monthly plans and directions written only in Japanese. I will give equal time to both in this chapter.

First, you must admit certain facts. If your child's list contains an item that he or she ABSOLUTELY POSITIVELY MUST HAVE, it will NEVER EVER BE ON SALE. Even if you scour eBay and find a "slightly used" (read: broken) model, the seller will still be demanding full price. Such was the case with my eldest daughter's cell phone.

When Natalie turned eleven and entered the hormonally charged educational tank known as "junior high," we began lengthening her leash, allowing her extra freedoms. A sleepover here, a parentless trip to the mall there, a football game here. Her friends enjoyed the same independence, which explained why cell phones—along with breasts—quickly sprouted from their bodies. Car pool backseats grew silent as her friends spent entire rides to volleyball practice tapping out messages, often to each other. That's right, it required satellites orbiting the earth to help my daughter communicate with friends that were sitting exactly three inches from her. But when your phone plan comes with unlimited texting, why not take advantage of everything NASA has to offer?

What seems like such a simple addition to a child's life—a phone—is, in reality, a frighteningly difficult decision for a parent. Dozens of questions filled our heads: Would Natalie use the phone responsibly? Would she use the obligatory cell phone camera to take photos of herself and her friends and post them on YouTube or Flickr, where they will live in infamy? And God forbid, would she become the next Alexa Longuiera?

On July 11, 2009, fifteen-year-old Alexa Longuiera was walking down a street in her Staten Island neighborhood, preparing to send a text message. The text was not coming from her phone but from a phone that a friend had just handed her.

That's right; she was with a *friend*, making the ensuing incident even more unbelievable (and hilarious if you are a comedian who has devoted the last fifteen years to writing jokes about technology).

As Ms. Longuiera continued to tap away at the phone, and her friend exhibited all the awareness of a hibernating bear, neither noticed an open manhole on the sidewalk. Suddenly Ms. Longuiera was gone.

Not gone like Amelia Earhart mind you, just temporarily gone. After plunging into the hole, and probably texting, "BTW, JUST FELL IN HOLE," she climbed out via a ladder that, up until July 11, 2009, had only been used by sewer workers and large rats.

She emerged relatively unharmed yet that didn't stop her parents from announcing to all media outlets that they planned to sue the city for not properly marking the manhole with a sign that, in all likelihood, should have said, "STOP TEXTING AND LOOK DOWN ALEXA!" I have been unable to find any information on whether a lawsuit has actually been filed but it sounds like a dispute that could easily be handled by Judge Judy.

In spite of the Alexa incident, one by one Natalie's friends received phones until my daughter was convinced she, save for cloistered nuns and Aborigine tribesmen, was the only person in the world without this communication necessity. She was left to peer over her friends' shoulders or worse, use my phone to send text messages. One day I looked at my address book and names like "Lauren," "Cheyenne," "Ali," and "Haley" had been added.

When the phone appeared on her Christmas list, we struck a deal: good grades = new phone. Seeing a light at the end of the tunnel, Natalie, already an excellent student, began requesting trips to every Best Buy and Sprint store within a fifty-mile radius of our home. Once inside, she pored over the phone selection the same way a gourmet chef looks at onions. Numerous phones were discarded for various reasons including "no slide-out keyboard," "not enough buttons," and "it only comes in red." The phones all looked the same to me but I wasn't eleven and therefore, had no idea what I was talking about.

Eventually she made her selection: a Sprint Rant. Oh, and she also informed us that it had to be purple. I'm not sure if purple phones get better reception but I could only imagine Alexander Graham Bell doing flip flops in his grave as he realized his history-making invention was being judged based on color.

At this point, it was time for deal number two: paying for it. We were determined to discuss this subject in detail as we had heard horror stories from neighbors who didn't properly explain what is and what is not included on a cell phone plan. As a result, they were greeted with thousand dollar bills for services such as "HANNAH MONTANA'S RINGTONE OF THE DAY!"

We agreed to buy the phone and add it to our family plan. Actually, *buying* a phone is relatively cheap. Some companies even offer to give you the phone for free providing you sign up for multi-year plans. In this age of rapidly changing technology, it amazes me that people would sign up for a multi-year *anything*. We purchased TiVo a few years ago and were offered a LIFETIME subscription. Seriously, does anyone think that in thirty years, our society will be using TiVo to record programs? By that time we should have the ability to reach into the television set, pull out the entire cast of *Seinfeld* and make them act out the show live in our living rooms. For a slight fee, of course.

Natalie's role was simple: chip in fifteen bucks every month. If she didn't have the money, we'd keep her phone for the entire month. Of course that would mean deciding whether we wanted to return text messages from Lauren, Cheyenne, Ali and Haley with probing questions such as "WHAT R U DOIN?"

With just a slight eye roll, she agreed to our terms and promptly retreated to her room to ponder a future career as a babysitter.

The fall whizzed by. Her grades kicked butt. She made the *High* Honor Roll. For some strange reason, her school has created an Honor Roll and a High Honor Roll. During awards ceremonies, I always feel sorry for the Honor Roll kids. Their certificates might as well say, "Congratulations on being slightly dumber than the kids sitting over there."

As Christmas approached Natalie went into cell phone training. By that I meant she began talking about cell phones with her friends the way an engineer talks about widgets. She boldly asked to see her friends' phones, comparing them to the Rant that she still did not have. She even sent us text messages from their phones, creating total confusion. When "CAN U PICK ME UP NOW?" appeared on my screen one afternoon, I didn't know if it was from my daughter or if it meant her friend Ali was running away from home.

Finally the blessed day arrived. December 25, the birthday of our Lord and the day my daughter became cool again. Sue decided to prolong the moment by wrapping the fully functioning Rant in eight boxes and dialing the number as Natalie unwrapped them. However, due to our cluelessness, we failed to maximize the volume so nobody heard the ringing phone as Natalie tore through box after empty box. Eventually the phone made its debut, with the same amount of enthusiasm as a newborn passing through a birth canal. Within minutes Natalie was texting her friends and actually getting responses, never mind that it was Christmas morning.

We had solved the budgetary issues associated with her cell phone. Or so we thought. What we didn't realize is that having a daughter with a phone leads to another equation: smart child + cell phone = not so smart child.

Now don't get all upset; I am not calling my High Honor Roll daughter dumb. However, there is no denying that her grammar and spelling skills suffered immediately as she chose to forego spoken communication and instead talk to her parents via a series of incomplete sentences, improper syntax, nonexistent punctuation and misspelled words. The last one grated on me because I have always prided myself on being an excellent speller and I have little use for letters and e-mails that are riddled with spelling errors, particularly when they come from college-educated adults who should, at the very least, know how to use a spell checker.

My appreciation of spelling is what attracts me each spring to one of my favorite sporting events, the Scripps National Spelling Bee. ABC televises the Bee just so we are annually reminded that we, as a nation, suck at spelling.

I first became hooked on the Scripps spelling bee in 2008, when it was part of ESPN programming. There are more ESPN channels than actual sports, which explains why I can see such raw athletic pursuits as Keg Tossing, Women's Bodybuilding and Horseshoe Pitching all from the privacy of my couch. Or I can watch documentaries such as *Legends of Cricket*.

On this particular evening, none of those programs interested me so I kept clicking until I came upon a bunch of junior-high-aged children, most with glasses, some with peach fuzz under their noses and all with pained expressions on their faces as they struggled to spell words that were never on any SAT test that I took. I never used any of them while writing for the *Palm Beach Post*. As far as I could tell, none of these words even *existed*. And yet these kids plowed through them as if they were plucked from a first-grade reading book.

I was transfixed. I marveled at the competitive nature and the drama. Yes, the Scripps spelling bee is pressure-packed, as evidenced in 2004 when thirteen-year-old Akshay Buddiga fainted onstage yet still returned to spell *alopecoid* correctly.

Of course my favorite part of the Bee is the fact that it has announcers. Since this event is deemed a sport, ABC must hire commentators to describe every scintillating SPELL-binding moment. These individuals are either new hires or were demoted after committing some grievous on-air violation like misidentifying the world champion keg tosser. Still, they announce the Bee with the same flair as Al Michaels describing the 1980 U.S. hockey victory over Russia:

Well, it's a beautiful day in Washington DC, Chip!

That's right, Jack. The air is thick with the smell of nouns, verbs and compound modifiers. We've got a top-notch bunch of kids who are going to stand on a stage with numbers around their necks and do nothing for most of the time. And we'll cover every step of the action.

You know, Jack, it's been said that American kids are too lazy to spell.

I disagree, Chip. I think Americans are excellent spellers, myself included.

Okay Jack, spell finite.

That's easy...F-I-N-E-K-N-I-G-H-T.

Let's go to a commercial, Jack.

According to Wikipedia, my favorite information source, Frank Neuhauser won the first Scripps National Spelling Bee in 1925, by correctly spelling *gladiolus*, which is either the center part of the sternum or a flowering plant having sword-shaped leaves. Upon reading that, I hope I never have to visit a hospital with *gladiolus* problems, as I don't want the doctor pulling plants from my backyard and saying, "They look okay to me!"

Through the years, some of the winning contestants seemed to get off easy. In 1934, Sarah Wilson won by spelling *deteriorating*. The following year Clara Moller correctly spelled *intelligible* to achieve Spelling Bee glory. Other winning words included *therapy, initials, vignette* and *condominium*. Melody Sachko spelled *condominium* in 1956, perhaps becoming the only champ to spell something that many people own at least once in their lives, unless you've chosen to be a *gladiolus* farmer.

The guy who got off the easiest had to be 1984 champion Daniel Greenblatt who won by spelling (are you ready?) *luge*! The winning word had four letters? That's like Tom Brady winning the Super Bowl by walking one hundred yards through the defense.

Spelling Bee organizers obviously had egg on their faces following Mr. Greenblatt's victory as the winning words got infinitely harder afterwards. In 1985, Balu Natarajan won for spelling *milieu*. It should also be noted that Balu Natarajan correctly spelled his own name, which I'm sure earned him extra points. Having a nearly unintelligible name seems to be a prerequisite to winning the bee, as Pratyush Buddiga (2002), Sai R. Gunturi (2003), Anurag Kashyap (2005), Kavya Shivashankar (2009) and Anamika Veeramani (2010) can attest.

In 2008, the year I discovered the Bee, twelve-year-old Sameer Mishra from Lafayette, Indiana triumphed. The young Mishra lad's fallen competitors included Sidharth Chand and Samia Nawaz. According to the National Spelling Bee website, Mishra plays the violin, enjoys computer games, likes math and science and aspires to be a neurologist. When I was twelve, I aspired to spell *aspire*.

Mishra blitzed the competition by spelling such commonly used words as *sinicize, hyphaeresis, taleggio, nacarat,* and *numnah*. Let me point out that the Microsoft spell-checker has ABSOLUTELY NO IDEA what any of those words mean as evidenced by the squiggly red line that appears under each word the moment I type it on my PC. I had to go to Encarta (the smarter Microsoft application) to learn that *sinicize* means "to acquire a Chinese idiom, form or cultural trait." *Hyphaeresis* is "the omission of a sound, letter, or syllable from a word." If you want a semi-soft cheese made from whole cow's milk, ask for some *taleggio*. If you wake up and your skin is a pale red color with a cast of orange, it may appear *nacarat*. Finally, the next time you ride a horse, make sure it has a *numnah*, a thick felt pad that prevents the saddle from moving.

My only beef with the National Spelling Bee is that these are words that, as I alluded to earlier, will never be used in normal conversation. I have been to Washington DC numerous times yet never felt compelled to approach a passing stranger and say, "Hello, my name is Greg. Can you please direct me to Chinatown? I'd like to *sinicize* before I leave for the airport."

Resigned to the fact that our cell-phone-toting daughter would never stand on stage in Washington DC with a number around her neck, Sue and I were forced to quickly learn cell phone lingo. This is among the dilemmas parents face while raising kids in the dizzying world of technology; kids learn to use this stuff because they *want* to, while parents learn because they have no choice.

Although I had better things to do, I eventually memorized some basic text. The aforementioned BTW meant "by the way." "R" was short for "are" or "our" and could easily be used multiple times in the same sentence. Example: R we eating at R house tonight?

Still, I was occasionally stumped. One afternoon I texted a simple question: "Can you clean the family room when you get home?" Note: Notice how I spelled out "you" instead of taking the shortcut and typing "U." Doing so, I felt, would cause somebody from Northwestern University's Medill School of Journalism to repossess my diploma.

Her reply? "K."

Nothing else. Just "K."

This is where the text conversation between father and daughter terminated. Drawing hieroglyphics on the walls of her room would be more effective than staring at a three-inch phone screen trying to decipher her answers. I elected to call her, knowing that she would most likely be NEAR THE PHONE.

"I just received 'K' on my phone," I said.

"Yeah?" she answered.

"Did you drop the phone? What is 'K'?"

"Dad, it's the abbreviation for *OK*," she responded, in her best "Obviously-my-Dad-never-made-the-High-Honor-Role" tone.

"*OK* is an abbreviation," I reminded her.

I decided then and there to fight fire with fire. A cryptic text from her would be followed by an equally cryptic text from me. One Friday evening I was reading in the living room, awaiting her arrival.

"You're late," I said, when she bounded through the door.

"No, I'm not," she replied. "I texted you. BBB10. Be back by ten."

"I got that," I said. "Didn't you get my response? HYBHTHBT?"

She looked at me quizzically.

"Have your butt home two hours before that."

That was definitely worth fifteen bucks.

I will now discuss the other way to purchase an ungodly expensive piece of technology on your child's wish list. Read on only if you are an insomniac and have a Black Belt in some form of martial arts. If you're still reading, you are a prime candidate for a Door Buster sale.

Earlier I mentioned that prized, high-demand electronics never go on sale. That is not *entirely* true. Door Busters, also known as Black Friday sales because they take place the day (I'm sorry, the ungodly early morning) following Thanksgiving, are the exception. These sales were invented solely because personnel from every retail establishment, including those that sell nothing but live bait, decided that sales figures for the entire year should hinge on the single day that follows gluttony, football and tense relations with relatives.

Door Buster sales also exist so television news crews have something to show on a slow news day. Invariably these "packages" (a term from my old TV reporting days) contain only images of fully grown adults acting like a combination of toddlers and gangbangers as they violently fight over whatever popular item the offending retailer chose to put on sale for 50 percent off just hours after the Thanksgiving dishes have been cleared away.

Occasionally this YouTube display of news turns into actual news; witness November 29, 2008, when security guard Jdimytai Damour was trampled TO DEATH at a Long Island Wal-Mart as customers surged forward to purchase, among other things, a $28 Bissel Compact Upright Vacuum. On that morning, Damour's first Black Friday job responsibility—and ultimately his last—was simply to *open the door.*

In spite of Damour's fate, and similar occurrences with slightly less horrific results (some shoppers merely suffer broken bones in exchange for a DVD player), retailers continue this macabre practice. In the event of mayhem, their savvy marketing departments cook up prepared statements that read with all the sincerity of those recited by professional athletes after being caught with steroids, handguns, stolen stereo equipment or all three.

We truly regret this tragic and unfortunate incident. We are cooperating with authorities and are confident that, in time, all the facts will come out. Until then, COME TO OUR EARLY BIRD 4 A.M. SALE! SIXTY-INCH FLAT-SCREEN PLASMA TELEVISIONS JUST $29.99. ONLY THREE IN STOCK!

On the day before Thanksgiving, my wife scours the ads—both print and online—to see if any Door Buster sale items match anything on our daughters' Christmas lists. Thankfully that had never been the case.

Until November 2009.

Having mastered the Wii games mentioned in Chapter Five, Amy requested something known as *Wii Fit*. I'm still not sure what it is, although the Wii website promised it would "improve balance, body mass index, and body control."

Too bad *Wii Fit* hadn't been invented a year earlier. If Door Buster shoppers had an ounce of body control, Mr. Damour might still be alive.

Normally $90, a store called Meijer had priced *Wii Fit* at $44.99 on *Thanksgiving morning*. That's right, Meijer, one of those stores with an identity crisis (groceries to the right, snow tires to the left, wool socks and Venetian blinds straight ahead) was having a Black Thursday sale beginning at 6:00 a.m. Would I wait in line and get one, Sue asked?

Until then the only time I had ever stood in line longer than thirty minutes for anything was 1981 when Bruce Springsteen's River Tour came through Chicago. I remember queuing up outside a record store four hours before tickets went on sale. Others ahead of me had obviously been there all night, judging by the sight of sleeping bags and scent of body odor. I spent the time chatting with fellow Springsteen fans, listening to his tunes, soaking in stories from Springsteen concert veterans and even sharing cheap wine from a hip flask.

I did score tickets that morning. Not great tickets mind you but tickets nonetheless. And the Boss did not disappoint. Twenty-eight years later, standing in line for something that improved body mass did not seem as appealing, even if I brought a nice dry Pinot Grigio to share.

Yet I succumbed to my wife's request with minimal complaining. Truth be known, I was looking forward to it. I'm an early riser by nature so the idea of setting a Thanksgiving Day alarm didn't seem that ludicrous. Besides, the store was only ten minutes away from my health club. What better way to begin Turkey Day than by making my daughter happy, saving fifty bucks, and squeezing in a five-mile run on the treadmill, thereby burning the calories in one scoop of mashed potatoes?

I awoke at 4:40 a.m. to the sound of freezing rain pelting my bedroom windows. This was no surprise; Murphy's Law specifically states that if one is going to wait outside a locked store for an inordinate amount of time, it MUST be raining, snowing, hailing or trembling from earthquake aftershocks. As I would soon find out, none of these calamities deter a Door Buster shopper.

I grabbed a sweatshirt, my North Face winter coat, a ski hat and gloves and pulled out of my driveway at 4:50, armed with nothing more than a cup of coffee and my Door Buster game face. As I journeyed toward Meijer, I saw other cars on the road. Suffice it to say that, if you are in your car at 5:00 a.m. on Thanksgiving, it can be for one of two reasons:

1) You are heading to a Door Buster sale
2) You need to dispose of a body...QUICKLY!

I live in a fairly safe neighborhood so I naturally assumed everybody who I passed or passed me fell into Category One. I also decided everybody was headed to Meijer in search of a *Wii Fit*, which made me press down a little more sharply on the gas pedal. I even cut off a few motorists, just to be safe.

At 5:05 a.m., I pulled into the Meijer parking lot, now three-quarters full with cars and one TV news truck. But where was the line? You know, the line of damp, sleepy customers preparing to trample the security guard? It did not exist. Instead, I saw people *entering* the store.

Did my wife misread the ad? Did Black Thursday actually start earlier than 6:00 a.m.? Had I failed before I even started?

Turns out, Meijer is open 24 hours so customers are free to come and go any time. But, as the ad promised, Black Thursday sales would not begin before 6:00 a.m. Customers could wait in line until then.

But which line? I sauntered to the electronics section at the store's rear to find about seventy-five people standing in a surprisingly orderly fashion.

"Is this the *Wii Fit* line?" I asked the last woman in line.

"No, this is the iPod Nano line," she replied.

"The *Wii Fit* line is two aisles over," said a Meijer employee, gesturing randomly with one hand while pushing a shopping cart full of merchandise with his other.

Immediately I saw one thing about this Meijer place that I liked, namely foresight to split up the lines as opposed to lumping everybody in a single mass. Plus, we were inside! This was going to be a good day!

I took a hard right, counted two aisles, took a left and almost tripped over a patron seated on the floor. I discovered this gentleman was "*Wii Fit* Door Buster Customer Number One" and, for all I know, had been sitting there since last Thanksgiving.

I followed the line down the aisle, which contained camping equipment. It made a gradual turn to the left and spilled over into the next aisle, which featured school supplies. Half-heartedly counting in my head, I estimated there to be about forty shoppers ahead of me. Judging from their body sizes all looked to be buying the *Wii Fit* for somebody other than themselves. Either that, or Wii Snack was also on sale.

I took a spot behind a woman who appeared to be about sixty. A fifty-something gentleman got in line behind me and the phalanx of *Wii Fit* hopefuls continued to grow. Within moments the line had increased by at least thirty. As it multiplied, a rough-looking couple trudged to the end. I heard the woman exclaim loudly to her partner, "Baby there ain't no way we're gonna get one of these fucking things."

I was thinking the same thing but chose not to express it publicly.

At 5:15 a.m., a Meijer manager appeared halfway through the line and announced, to no one in particular, that the store only had twenty *Wii Fits*.

"You're welcome to wait but I'm just telling you what we have," he said, before disappearing into the stock room, presumably for his own protection.

At this point, my predicament read like a second grade math problem: *You are the forty-first person in line for a toy. A grown up says there are only twenty toys available. Will you get a toy? Please show all work.*

Common sense dictated that I should get out of line. But, upon hearing the employee's grim news, exactly ZERO people moved from their places, including Ms. Potty Mouth well behind me.

These people must know something I don't, I thought. If they're not moving, I'm not moving.

Door Buster shoppers are, if nothing else, eternally optimistic. I could almost hear them rationalizing how a *Wii Fit* could still be theirs.

Maybe at least ten people in front of me will all have fatal heart attacks in the next forty-five minutes, their faces appeared to say.

Or maybe ten would get trampled once the clock struck six. I decided to wait.

A few minutes later the same Meijer employee appeared and announced that the store actually had twenty-nine *Wii Fits* available "and some *Wii Fit Pluses*." The *Wii Fit Plus*, by the way, is a slightly more expensive BUT STILL 50 PERCENT OFF ON DOOR BUSTER THURSDAY AT MEIJER model.

This was the first time I had ever heard of a store suddenly discovering MORE merchandise. Whenever I go shopping at the mall and ask if the store contains a particular item in my size, the response invariably is, "That's all we have." Nobody has ever said, "You need that in a large? Hang on; I think a truckload of larges just came in. I will go get one for you because I am a dedicated store employee."

By now I realized there was no rhyme or reason to a Door Buster sale. Twenty *Wii Fits* had just become twenty-nine. The ever-optimistic shoppers were even more jovial, assuming that twenty-nine would soon turn into sixty, maybe more. Perhaps they would come with free LIFETIME warranties! Even the guy behind me, who had donned and removed his coat at least three times in forty-five minutes, took it off again as if to say, "I'm in it for the long haul as well." We began to bond as only males who have been sent to Door Buster sales by their wives can do.

"If I get the last one, I promise you can come over and play with it any time," I said.

He chuckled and said he'd take me up on it.

At 5:59 a.m., the line was filled with the same kind of anticipation normally reserved for male Mardi Gras revelers as a woman prepares to display her breasts. At 6:03 a.m., the line began moving. I moved out of the school supplies aisle, around the corner and re-entered the camping aisle. I noticed a store end cap containing a display of hunting knives. Bad idea, I thought, to let aggressive, overcaffeinated Black Thursday shoppers anywhere near weapons.

From down the aisle, out of my line of vision but within earshot, came the first Black Thursday argument. I'm not sure what it was about but a clearly agitated woman kept saying, "I want my receipt and I want it NOW!"

Upon hearing her screams, the TV news crew scrambled into position.

At 6:13 a.m., the Meijer employee delivered the worst news I've heard since the Cubs signed Milton Bradley: only two *Wii Fits* remained.

This time I did an exact count of customers in front of me rather than an estimate. There were eleven patrons, none of whom moved in spite of the simple math equation: 11 desperate shoppers – 2 *Wii Fits* = 9 losers.

It was time to get out of line. My compatriot behind me put on his coat for the umpteenth time and did not take it off. Instead, he followed me down the aisle toward the exit, muttering something about "a perfectly good day wasted." He would soon realize that the sun had not yet risen over the horizon. Technically it was still nighttime.

I exited the store and strode to my car, where my gym bag awaited. On this Thanksgiving morning I was thankful that, in spite of the horrific economy, paying regular price for a *Wii Fit* wouldn't break the Schwem bank account.

I turned on the radio. Bruce Springsteen was singing, "Santa Claus is Coming to Town." I pulled out my cell phone and texted my wife:

NO WII. 2 MANY PPL.

Oh, the irony.

CHAPTER 7

What's on YOUR Mind?

The best thing about Wiis and cell phones is both allow you to disable the Internet feature, meaning you can breathe easy knowing your children are entertaining themselves without inadvertently stumbling into a chat room occupied by a man whose screen name is "Taylor Lautner" but in reality is "Thomas, the Wii-playing, cell-phone-toting sicko."

The *real* problems come when you, as a parent, must evaluate entertainment forms that are exclusively powered by the Internet.

Enter Facebook.

One day I found myself waiting out an excruciatingly long layover in the Lehigh, Pennsylvania, airport (proving once again that the longer the delay, the smaller the airport) when I opened my laptop and typed the following sentences:

I am typing

I am in Lehigh, Pennsylvania, home to Amish people.

I am looking for Amish people but can't find any.

I am taking a sip from the Diet Coke that I purchased at the airport Subway. It is cold and refreshing.

I am checking my Blackberry to see if I have any e-mails.

I think I just saw an Amish guy.

What is an Amish guy doing at an airport?

If you type sentences like this, you either have some form of obsessive compulsive disorder or you are a member of Facebook, the wildly popular social networking site that invites its members to announce to the cyberspace community exactly what is on their minds and what they are doing RIGHT NOW!

You know what's on my mind right now? Two questions:

1) Should I let my children join Facebook?
2) Why did I join Facebook?

Deciding whether you should let your kids on Facebook is kind of like wondering how you'll respond when they ask if you've ever smoked marijuana. Do you lie and say, "No," or do you reply, "Yes, but YOU should stay away from it."

At last count, Facebook was approaching McDonald's status in terms of number of customers. McDonald's used to actually count the patrons who passed underneath its golden arches to get a Big-Mac-and-fry-fix. Their marketing department would gleefully change the signs every so often; "Over 25 billion served" would magically become "Over 26 billion served." Whoever was in charge of sign changing at McDonald's obviously got a pink slip the day the franchise decided to go with "billions and billions served."

The most common description I read about Facebook is that, if it were a country, it would be the fourth largest in the world! I'm not sure how residents of Indonesia, currently the fourth largest country, feel about this but chances are they are discussing it on their Facebook walls.

What started as a quirky idea in a Harvard dorm room has snowballed into a phenomenon that, in my opinion, threatens to overtake Fantasy Football as the biggest time waster in modern history. And much like karaoke, it's not going away anytime soon.

As of now, my children are immune to Facebook but I know the day is coming when they will set up their own profiles, upload photos taken at parties and then, ten years later, wonder why they can't get jobs.

Mind you, Facebook is not the first social networking site to hit the Internet. MySpace preceded it as did LinkedIn, which I joined several years ago on the advice of a friend who termed it a "business-oriented" social networking site. By "business-oriented" it means that the members actually have jobs and, furthermore, actual lives.

Such does not appear to be the case with Facebook.

I took the Facebook plunge only after a marketing executive told me it would increase my online profile and allow people trolling cyberspace one more way to reach me directly. I took the bait even though I feel the *last* thing we need is another method of contact. Nobody home when you knock on my door? No problem.

Just text me, e-mail me, fax me, ping me, IM me, call me, nudge me or poke me. If I still don't answer, check the morgue.

I went to the Facebook site and set up my free profile. This took just over ninety-two hours because Facebook wanted to know EVERYTHING about me, including things I've never told my wife. Oh sure, the questions start simple enough: Was I married? Single? Engaged? In an open relationship? Or my favorite choice: "It's complicated." Excuse me, O Facebook Gods, but what relationship IS NOT complicated?

Was I interested in men or women? Actually I find both men and women interesting, particularly if they've had a few drinks and, like me, are in the midst of a lengthy airport delay. Or in a hot tub. Something about a hot tub makes people chatty though I can't figure out why. But I assumed Facebook wanted to know my sexual preference. I left it blank.

What were my political and religious views? I hoped there might be drop-down boxes to choose from. Had the choices included "Democrat but Sarah Palin makes me laugh" or "Church on Sundays unless my daughter has gymnastics," I would have made a choice. But I left those blank too simply because I didn't want to be contacted by any Facebook members with conversion on their minds.

The interrogation continued. What were my favorite movies? TV shows? Musical groups? Quotes?

The quotes box stumped me, as I have never gone through life quoting anybody other than my father who still lives by the mantra, "So help me God, I am turning this car around right now!" Somehow that didn't seem appropriate considering some of my

Facebook "friends" were quoting Plato, Sun Tzu, Lou Holtz and Lee Iacocca.

Yes, let's talk about friends.

If Dr. Seuss were still alive—and had joined Facebook—no doubt he would have written, *Oh the Friends You'll Meet!* instead of *Oh the Places You'll Go!* Facebook doesn't require you to go anywhere; everybody comes to you. Once I had completed my profile and announced every known fact about myself except what I ate for breakfast on June 29, 1981, (Note: Facebook support personnel are working to answer that question right now.) it was time to sit back and hear from others in the Facebook community who wanted me to be their friends.

It didn't take long.

Facebook, you see, crawls into your inner being and just keeps digging deeper, much like a tapeworm. Facebook can scan your e-mail address book and determine which contacts also have Facebook profiles. It will even contact them directly. If "Bob Smith" is in your address book, Facebook will send a message to more than one million Bob Smiths in the world, letting all of them know that you are now on Facebook and ready to become friends. All Bob(s) has to do is ask.

That's another thing about Facebook that I found humiliating: *asking* people for friendship. The closest I had come to this was in 1992 when I asked Sue to marry me. Asking people in cyberspace to friend you is sort of like arranging a play date for yourself.

And yet this strange ritual didn't seem to deter hundreds of Facebook users, most of whom I'd long forgotten, from requesting that I be their friend. There was a Russian woman who hired me ten years ago to perform for a company that folded during the market meltdown; a fellow Chicago comedian whose name and face I could barely place; and a Spokane resident who said we had been good friends in first grade. I took his word for it even though I have no memories of even *attending* first grade.

My sister Julie asked if I would be her friend. My SISTER! I was being demoted from blood relative to friend. I was so incensed I wrote back: "Are you the Julie that sat across the dinner table from me, had a bedroom down the hall, and shared the back seat on car trips? If so, then yes, I'll be your friend. But send me a photo first so I can be sure. Preferably topless."

My accountant also wanted to be my friend. Apparently he isn't satisfied with the monthly checks I send him.

I wanted to e-mail all these people back and say, "Where were you when I was seven, huh?" Nobody wanted to be friends with me then. There was no "Greg's a pretty cool kid, so let's ask him to play baseball with us" group that I could join. But with Facebook, suddenly I was more popular than I had ever been in my life. So I responded to everyone's friend requests in the affirmative, sometimes including a harmless message like, "What's up? It's been years!"

One great thing about Facebook is how quickly your ego swells. Even the nerdiest people can find friends. Better yet, they can find entire collections of nerds who share their interests by joining Facebook groups and fan pages. I would explore these groups shortly but first I had to invite all my newfound friends to write on my Facebook "wall."

Ten minutes after doing so, a favorite quote popped into my head:

"Big mistake. HUGE." —Julia Roberts, *Pretty Woman*

Opening up your Facebook wall to your friends is another way of trying to scare ants away from a picnic by pouring sugar on the ground. It allows them to, indeed, announce what they are doing right now and what is on their minds. Suddenly I was bombarded with "posts." A Chicago friend was "thrilled that the Sox pulled it out." A friend in LA was "working on a plan." A Denver friend "got sunburned during the kickball tournament" and a Long Island friend was "counting the minutes until she sees her boys on Wednesday." Because these messages are so brief, it's easy to let your imagination run wild when trying to decipher what they really mean. For all I knew, my Long Island friend's boys had been kidnapped and would be returned Wednesday, once the perpetrators had received the ransom money.

Approximately fifty of my friends were either professional meteorologists or desperately wanted to be. I made this deduction from the following posts:

Sigh…did we have summer?

Another beautiful day in Northern Illinois.

I'm so ready for sweater weather.

Gorgeous day in Chicago!

Love fall weather even when it happens in the summer.

Oooh it's hot today.

Beautiful morning in Cleveland.

Bear in mind that these friends actually sat at their PCs and *thought* about what they planned to write. This was the best they could come up with.

Of course, if I wasn't creative enough to peer out my window and write about the climate, other friends suggested verbiage for me to post. Like this one:

No one should die because they cannot afford health care, and no one should go broke because they get sick. If you agree, please post this as your status for the rest of the day.

Although I agree with that statement, I neglected to post it—probably for the same reason that I don't have a "HONK IF YOU'RE HORNY" bumper sticker on my car.

Before Facebook launched its current look, it gave users a drop-down menu containing "What are you doing right now?" choices. So, if a user wasn't actually sure what he or she was doing right now, Facebook could help. One of the choices was "going to bed." That's right, I could let my entire collection of Facebook friends know that I was "going to bed" with a simple mouseclick. Trust me, if anybody interrupted my life with a post stating they were going to bed, I would tell Facebook to WAKE THEM UP. IMMEDIATELY!

Now that I had set up my own profile and explored the basics of Facebook, I was ready to begin my original quest—determining if Facebook was appropriate for my children. The quickest way to do this, I decided, was to check out Facebook groups.

If Facebook itself is the fourth largest country, the number of groups created by its members would comprise the fifth largest. Many of these groups promise that something wonderful will happen just by joining. A few examples: "IF 5 MILLION PEOPLE JOIN, MY MOM WILL STOP SMOKING," "I BET I CAN FIND ONE MILLION PEOPLE WHO HATE MOSQUITOES," and my favorite, "IF ONE MILLON PEOPLE JOIN, MY DAD WILL NOT PUT OUR DOG TO SLEEP." Between Christmas Eve and December 26, 2009, that group's ranks had swelled from twelve thousand to two hundred thousand. By January 5, 2010, more than 1.4 million Facebook participants had ensured the dog would continue living. Many members suggested the dad be put to sleep instead.

Although I wanted to view Facebook groups that might attract my kids' interest, my selfish impulses took over. Since I own my own company, I chose the "business" category and was immediately bombarded with more than five hundred groups. The first group, containing 530,000 members was titled, "PEOPLE WHO ALWAYS HAVE TO SPELL THEIR NAMES FOR OTHER PEOPLE." Not certain what this had to do with business, I clicked and learned it was actually a group for people "with an unusual spelling or an unusual name." I clicked "view members" and realized how happy people like Aoife Nic An Tsaoir, Zsuzsanna Göröncsér, Justyna Magdalena Przeworek and Monsef Zoufir must have been upon realizing they were not alone. I contacted all of them and suggested they enter the National Spelling Bee. I also wondered why Becky Gordon was in the group.

There were 44,481 Facebook members who had joined "I HAVE WORKED IN RETAIL AND THUS HAVE LOST ALL FAITH IN HUMANITY." Upon seeing this I was tempted to start my own group entitled, "I HAVE JUST BEEN TO WAL-MART

AND HAVE THUS LOST ALL FAITH IN ANYBODY WHO WORKS IN RETAIL."

A surprisingly large number of business groups had titles and descriptions written in Arabic or something that looked like Arabic. The paranoid part of me began wondering if terrorists were communicating via Facebook but I chose to put that ugly thought out of my head. Instead, I perused, "WE WANT STARBUCKS IN NORWAY." I joined that one because, if I ever find myself in Norway, it might be kind of fun to order a "tall half-skinny half-one-percent extra hot split quad shot latte" from a barista who only speaks Norwegian.

Speaking of foreign countries, another group that fell into the "business" category was "HOW TO TREAT BARSTAFF IN THE UK." Judging by the 13,705 Facebook members who had joined, this is a serious issue. I've been to England a few times and don't remember any uncomfortable encounters with bar staff. I merely ordered a pint, put money on the bar and that was that. But, according to one member, the correct way to treat UK bar staff is, "Mouth closed, money out! Just because we look at you, doesn't mean we're ready for you! Just because we haven't looked at you doesn't mean that we don't know you are there."

That may be true in the UK but it certainly doesn't apply to Irish bars in the States, a fact I learned in summer 1994 when the World Cup soccer tournament came to America. While walking through Greenwich Village one sweltering evening, I popped into an Irish bar, hoping to soak in the atmosphere. To say the place was up for grabs is a gross understatement; I've seen Iranian protests on CNN that were more peaceful.

Dozens of young men, dressed in the exact same uniforms worn by the Irish national team, were singing native songs, violently chest-bumping one another and downing pints of whatever came out of the tap at the rate of one every twelve seconds. Eventually I made my way to the bar but not before getting elbowed several times in the ribs and once just under my left nostril. I stood there for at least twenty minutes while the bartender repeatedly glanced at me while serving every other patron—twice.

Eventually I managed to get a beer. Pinned to the bar and unable to move, I focused on the mosh pit that had formed near the pool table when another patron yelled in my ear.

"Here for the Cup, are ye?" I believe he bellowed.

"No, I'm here for a business conference," I yelled in return. Then, because I couldn't think of anything dumber to say, and because the only thing I know about World Cup is that it's played with a soccer ball, I screamed, "I assume Ireland won today?"

"Nah, Ireland didn't even play today," he replied before climbing on his stool and swan diving into the crowd.

Since that day, whenever Ireland *wins* a soccer game, I immediately hide under my couch for at least forty-eight hours.

I could have spent days reading, joining, and leaving assorted Facebook business groups but I remembered my mission: determine if Facebook was suitable for my children. So far it seemed harmless enough—much less damaging than Spring Break in Fort Lauderdale or South Padre Island, Texas. I've seen the *Girls Gone Wild* videos.

But I decided to peruse "Student Groups" just to be certain. I further narrowed it down to "Student Groups/Social Groups" and dove in.

Oh, the choices my kids will have once they get Facebook accounts. Would they sign up for "ONE MILLION PEOPLE FOR A TITTIES BUTTON" or "I BET I CAN FIND A MILLION MEMBERS WHO HAVE GOT WASTED IN A STUDENT UNION?" How about "102 THINGS GIRLS NEED TO KNOW ABOUT GUYS." Number four, by the way, is "If he slapped you hard, you probably deserve it." Girls, if you don't agree with that statement, you will also be devastated to learn that this group boasts 413 members while "HELP STOP WORLD HUNGER" lags behind with 184.

On that note, I have always wanted my kids to be socially aware and fight for causes they believe in. The more I perused Facebook groups, the more I thought that it might be a perfect platform to find a cause that interested them, titties buttons notwithstanding. Granted, some causes seemed a little ridiculous; one hundred members had joined "TEENAGERS AGAINST CANCER IN BAHRAIN." I'm sure they are a noble, dedicated bunch but have you ever met anyone who is for cancer? I'd rather my daughters devote their attention to "STOP MUTILATION TO FURTHER EDUCATION," or a group that has already found a solution to its cause. A group like "STOP HITTING ON YOUR PROFESSOR. IT WON'T CHANGE YOUR GRADE."

Maybe I have nothing to worry about. Recent studies show Facebook is aging, with the over 55 set comprising the fastest growing demographic. But my kids will eventually find Facebook, or some permutation thereof. And that means their lives have the potential to become open books, read by anybody. And, if they upload photos or movie clips, viewed by anybody. There is also the potential for them to meet and chat with total strangers, something that I discovered young people are prone to do. How did I discover this? By contacting them of course!

BULLETIN: I AM NOT, REPEAT…NOT! A! PERVERT! Read on.

The reason for my inquiries was a news item that made the national wires in November 2009. Apparently a fifteen-year-old boy in Buffalo Grove, Illinois, dialed 911 after his mother took away his Xbox video game. Nothing much came of the 911 call—the kid hung up mid-conversation—but the police called back and delivered a stern lecture on the importance of listening to one's parents.

I thought the boy and his mom might make a good chapter for this book so I set out to find them. Because the Xbox addict was fifteen, his name was not listed in any news articles. I placed a call to the Buffalo Grove Police Department where a media relations officer confirmed that yes, the boy lived in that department's jurisdiction. I explained my intentions and the officer agreed to contact the family and ask if they would talk to me. Two weeks went by and I had heard nothing so I elected another tactic: contact some Buffalo Grove High School kids.

Sure enough, a Facebook search revealed the existence of the BUFFALO GROVE HIGH SCHOOL CLASS 2010 group. I picked a few names at random and sent the following message:

Hi (name of kid). My name is Greg Schwem and I am a Chicago-based writer and professional stand-up comedian. I am looking for somebody who may be a fellow classmate of yours. There was a story in the Chicago Tribune last month about a fifteen-year-old Buffalo Grove boy who called 911 because his mom took away his Xbox. The story is attached. Are you or your classmates familiar with this story? If so, do you know this kid? I would like to interview him for a book I am working on. It would be illegal to use his name in print so I don't plan to do so. I just want to hear from him...and probably his mom too. Please read the article and reply to me either way. Ask your friends if they know him. Thanks. Hope your studies are going well. By the way, I am a 1980 Prospect High School graduate.

Within hours I had received several responses.

Yes, I am familiar with this story, but sadly I don't know who the boy is.

Never heard of that story but it's really funny.

I'm not familiar with this story... I will definitely ask around and let you know if I find anything out. These boys are crazy about their Xbox, especially with Call of Duty recently coming out! Good luck!!

Maybe these kids responded because they were trying to be helpful. But I still found it weird that they eagerly replied to a man who admitted he had graduated high school thirty years ago.

I'd still like to talk to the Xbox kid. But until he calls, I have formed my own Facebook group:

"I'LL BET I CAN FIND ONE MILLION TEENAGERS WHO WILL CONTACT A 47-YEAR-OLD MAN!"

CHAPTER 8

Twitter Me This, Coach

Having mastered Facebook and Wii, meaning I now knew the definition of each, I decided to ramp up my technology experimentation. Truth is, I was feeling pretty confident, much the same way I feel when I successfully complete a hardware-related project around my house, such as changing a lightbulb.

I anxiously awaited the next "killer app" just so I could play with it, use it effectively, learn the do's and don'ts and teach them to my kids when they discovered it.

Enter Twitter.

Twitter exploded on the scene in 2007, nearly one year after its creators, Biz Stone and Evan Williams, saw the need to create yet another form of communication because e-mail, voice mail, texting, videoconferencing, faxing, answering machines and yelling "hey" out the window weren't enough.

As far as I'm concerned, Twitter does for face-to-face communication what Bill Clinton did for Al Gore's 2000 presidential campaign—destroys it. Thanks to Twitter, humans can now speak to each other in 140-character messages. How the developers arrived at 140 characters is beyond me. I ponder that number in the same way that I wonder why an erection lasting more than four hours is cause for alarm. Who decided that four hours was the drop-dead time to reach for a phone and dial 911? Or Twitter 911, if that's possible. I have no medical training but am fairly certain a three-and-a-half-hour erection is far from normal.

Anybody using Twitter quickly realizes that 140 characters is very brief. To wit, the sentence you just read was seventy-three characters, counting spaces. And yes, Twitter does count spaces. Which is why most "tweets," as the messages are called, are chock full of useful information such as, "Just ordered Starbucks," followed by "Starbucks tastes good," and "Leaving Starbucks now."

I've never been good at condensing my thoughts, preferring to write—and speak—in free-flowing style. One of my first Northwestern journalism assignments was to write a 5,000 word article. This was before the days of word processing programs with features such as "word count," forcing me to hand count each word at 2:45 a.m. and, upon realizing that I was 567 words short, rephrase simple sentences such as "the man jumped in the car" to "the male species propelled himself in a forward motion toward the automotive vehicle." That's been my style ever since.

But seemingly overnight, EVERYBODY was tweeting. Celebrities tweeted, CNN anchors tweeted, politicians tweeted. In summer 2008, presidential candidate Barack Obama actually tweeted the selection of Joe Biden as his running mate. However, since Biden had not signed up for a Twitter account, he didn't hear the news until a week later.

Soon tweets were retweeted to bloggers who retweeted the retweets until eventually the world fell silent for just a moment while the entire universe tweeted.

With all these people tweeting, I assumed it would only be a matter of time before my children discovered this alternative method of speech. Plus, I was starting to wonder whether to believe the advice of social networking "experts," who were convinced my bookings would increase, my popularity would soar and my erection might last up to EIGHT hours if I became a member of the Twitter universe. Begrudgingly, I logged on to Twitter and established an account.

Thankfully, obtaining a Twitter account is infinitely less taxing than signing up for Facebook. It doesn't ask for personal information; the only thing Twitter cares about is whether or not you own a cell phone. Which is why Twitter is a great service for boring individuals.

Once online, I immediately became depressed. Whom would I tweet? Who would want to "follow" me and hang on my every tweet? If I tweeted "Just went through the carwash," would somebody tweet back, "How does the car look now? Did they miss a spot? TELL ME MORE!"

I stared at my Twitter homepage. There was an empty box staring back at me; a box anxiously awaiting my first tweet.

I hovered over the keyboard and wrote, after much thought, "Just signed up for Twitter."

That ought to draw some interest in Twitterland. And it was definitely better than, "It's cold today."

Alas, after two hours, I had received exactly zero e-mails from Twitter requesting my approval for anyone to follow me. Feeling a little like the kid picked last for the neighborhood basketball game, I messaged a friend via Facebook, where at least I had established a couple of friends.

"Are you on Twitter?" I asked.

"Yes, but I don't get it" came the reply from Janis, a Canadian business acquaintance. By "get it," she meant that she didn't understand the entire purpose of Twitter.

"Can you follow me?" I begged. It was like a girl asking a boy to take her to prom.

"Sure," she replied. "I want to see how this thing works."

"I'll do the same," I said, meaning I would follow her. Might as well make somebody else happy.

Moments later, my inbox roared to life: "Janis has requested to follow you on Twitter."

I eagerly accepted and composed my second tweet, this one a direct message to Janis: "Thanks for following me."

For a week, that was all the tweeting I did. In the meantime, I immersed myself in the Twitter lingo, learning that "@" meant I was tweeting a particular individual; "#" was a hash tag intended to group tweets into popular subjects; "RT" was short for retweet and

"@#!$%@#" meant I was using profanity while trying to figure out why the @#!$%@# I signed up for Twitter in the first place

When I calmed down, I signed up to follow a few people and media outlets including the *Chicago Tribune* (@chicagotribune) and CNN Breaking News (@cnnbrk). Apparently I signed up for Twitter during the slowest news period of the century for I received exactly one BREAKING NEWS tweet from CNN the first month and it concerned a country I had never heard of. More breaking news occurred via the *Tribune*, if one considers a Cubs victory breaking news.

I was ready to give up Twitter because it was depressing me. Not depressing in the sense that I had no followers save Janis; not depressing because I was getting *Tribune* tweets like "Man kills entire family in suburban Chicago home," but depressing because I had tweeted nothing. Could my life really be so uneventful that it wasn't even worth 140 characters? I've seen Britney Spears interviewed several times and she strikes me as somewhat boring, not to mention amazingly stupid. Yet she has about two jillion Twitter followers.

Then, while reading *USA Today* one morning, I happened on an article in the sports section—an article that focused on the use of Twitter by college and professional coaches.

It seems that coaches were tweeting fans with practice updates, tweeting boosters on blue chip signings and tweeting recruits and begging them to attend their respective institutions. That last one is probably illegal but I seriously doubt the NCAA has gotten around to hiring a Twitter cop.

Now here was something I could tweet about! For when I read the article, I was one month into a coaching assignment of my own. It was April and I was presiding over the Wildcats, a dozen of the cutest six- and seven-year-old girls in the local Little League "Kittens" division. Our first game was rapidly approaching. Could I handle managerial duties while tweeting at the same time? More importantly could I capture the thrills and excitement of a league where a "team mom" was a necessity and my coaching duties included formulating a "snack schedule?"

There was only one way to find out. I hooked my cell phone onto my belt, gathered up the softball equipment and prepared to introduce my Twitter followers (which now numbered thirteen) to the non-stop action of Kittens softball.

I neglected to tell the girls or their parents that I would be tweeting. Of course, seeing a parent at a Little League game tapping on a cell phone is very commonplace these days. I once witnessed a coach operating a pitching machine and sending text messages BETWEEN PITCHES! If the opposing team had inserted Derek Jeter into the lineup, Coach Multitasker would never have noticed.

I vowed to tweet discretely, and often. The following is the entire Twitter conversation:

1:00 p.m. Overcast and 75. The Wildcats are ready to play softball. The snack has arrived.

1:01 p.m. Amy just announced that she doesn't want to play catcher.

1:03 p.m. The Wildcats take the field. Coach Greg has put the "no cartwheel" rule into effect.

1:04 p.m. First question for Coach Greg: "Where is right field?" Didn't we cover that?

1:09 p.m. 1-0 Wildcats over Falcons. The girls said we just scored a "point."

1:12 p.m. Second inning. Kylie can't find her bat. Mikaela has lost her glove. My sanity is around here somewhere.

1:15 p.m. 3-2 Wildcats. Grace says she is "freezing." The temp has dropped to 72.

1:20 p.m. I don't feel comfortable helping a girl of ANY age put on a chest protector.

1:22 p.m. An empty baggie just fell from my pocket. A parent said, "Coach, you dropped your weed." I like this guy.

1:27 p.m. Just realized we neglected to cover the difference between "dropping" the bat and "throwing" the bat.

1:34 p.m. 6-4 Wildcats after 3. Our 3rd baseman just stepped on 3rd for a force. One problem...nobody was on base.

1:36 p.m. Ali takes first potty break of the season. Coach Greg doesn't have that luxury.

1:42 p.m. Elizabeth tagged a runner! The correct runner!

1:43 p.m. There's a big hole in center and there will be until Ali returns from the bathroom.

1:46 p.m. Falcons just scored. Holy @#!$%@#!

2:14 p.m. Coaches just realized the catcher is crying. Tough to see when she is wearing a mask.

2:15 p.m. Amy is the new catcher. More crying!

2:25 p.m. 6-5 Wildcats heading to the last inning. The girls are eyeing the snacks.

2:28 p.m. The girls are getting good at staring at the ball while it rolls past them.

2:29 p.m. Falcons on first and second with one out. GULP!

2:30 p.m. Grace just caught a pop-up, stepped on second and tagged a runner. That's four outs, correct?

2:34 p.m. Game over. We win. Juice box never tasted so good.

2:36 p.m. The Wildcats are 1-0. One victory and zero icepacks or injuries that drew blood. So far, a good season.

2:37 p.m. Almost forgot. Final score: 6 points to 5.

LIFE LESSONS

CHAPTER 9

Adventures of an Overseas Traveler

Almost any parent will tell you that it is tough traveling on business. I say "almost" because I have met parents who, I am sure, count the days until they can get away from the little delinquents that they unfortunately spawned.

While I enjoy the travel that comes with my job, I prefer it in small time increments. Usually I am gone for slightly more than twenty-four hours, enough time to blow into a hotel that morning or the previous evening, perform stand-up comedy at some point, and then hop a plane back to Chicago.

So naturally I was hesitant in May 2009 when I was invited to join a Sarasota, Florida-based insurance company on a trip through Europe.

For eight days.

For the insurance folks, this was an incentive trip, the kind that, depending on whom you ask, is "necessary" or "excessive." However, this company (which shall remain nameless simply because I liked its employees) was privately owned, did not accept TARP bailout funds and therefore operated on a "we will do as we please and we will have fun doing it" principle.

They would be socializing, drinking and networking. They would start in Amsterdam, end near Frankfurt and cruise the Mosel Valley, stopping along the way in towns I've never heard of including Xanten, Cochem and Treer. I would perform stand-up comedy, assist with the company's awards program (not an easy task aboard a moving ship, as I found out later) and, in general, keep these top salespeople and their spouses laughing and entertained for five days.

Tack on three days for travel and you've got eight days.

Alone.

On business.

I took an eight-day trip in 2004 sans kids. But Sue was with me and the trip was entirely pleasure. We left the kids with my in-laws, who possess the best babysitting item ever invented—better than a nanny.

An inground pool.

Trust me, when your kids are small, they can spend an entire summer wearing two articles of clothing: diapers and bathing suits.

Now my kids have schedules that would astound some CEOs. They go to SCHOOL, then they have ACTIVITIES and then they have MORE ACTIVITIES. And don't forget HOMEWORK!

I don't care if your kids have the disposition of a Golden Retriever seeing-eye dog and your relatives the patience of Job, eight days of playing "who needs to be where and when" and "I need help with this math problem" can have lasting consequences. Lawyers are summoned and wills are revised around Day Six.

So, even though the insurance people graciously invited my wife, we decided I would go to Europe solo while she stayed home in her "office," another word for "the car." It would be the longest I had ever been away from my family and, as I found out, a true test of communication skills.

Once again, we turned to technology to solve our dilemma. How to correspond? I briefly considered Skype, a video messaging service which allows you not only to talk with others, but also *see* them courtesy of a webcam that is needed to experience its true benefits. Many laptops have built-in webcams while others require you to mount a Star-Trek-looking eyeball contraption on your monitor. The problem with Skype is you actually have to *know ahead of time* when a party is calling so you can have your PC turned on and all your "skype-italia" in place. In our situation, the problem was also the time difference. One party would be talking in the early morning which, let's face it, is not a great time to appear on a webcam. Even Tom Cruise would look sort of gross.

Instead, we elected to correspond via overseas phone calls. A quick call to Sprint revealed that my ever-present Blackberry, which plays music, takes photos, surfs the Web and can connect to YouTube in an instant, would NOT work in Europe. The Sprint salesperson suggested I purchase a cheap international phone online.

"Have you ever heard of a website called eBay?" he semi-whispered, as if eBay were a secret, CIA interrogation camp.

Eventually I rented an international phone—over the phone—from a company called CellHire. Helpful salesperson Mike informed me that the phone would come with two SIM cards (another phrase I had never heard of), one for the Netherlands and one for Germany. All I had to do was swap out the cards, depending on what country I was in, and I could easily dial home—for eighty cents a minute.

When I'm on the road, calls home can often last upwards of thirty minutes by the time I have talked to all three family members. And that's if everybody had a good day. "Bad day" calls can last twice as long. The recession forced the Schwems, like most families, to make sacrifices when possible. We ate out less, drove less and socialized less. The sickly economy also forced us to talk less. We agreed that I would call home every other day.

That would be four calls over eight days.

I suggest that nobody ever use this formula. Instead, call whenever the mood strikes, costs be damned. Cancel HBO, take a second job or refinance the house if you have to.

Family communication, as I found out, is not something that should be *scheduled.* Establishing a timetable ensures the risk of calling on days when there is nothing to say and being out of touch on days when the sound of a family member's voice could lift your spirits exponentially.

To demonstrate, I will divide an eight-day business trip into Days Zero to Eight, using the "every other day" calling pattern, and will try and explain what occurs on each day.

An overseas trip doesn't begin on Day One; it actually starts on Day Zero. This is your first phone call, known simply as the "I am here" call. This is the briefest call, because, as I mentioned, you're calling the family when it's your morning and their night, or vice versa, and that's totally weird to both parties at this point.

So the call lasts about three minutes: "I'm here, how are you, how are the kids, I miss you already, yada yada yada."

You don't talk to the kids on this call unless they answer. You talk to your wife. Before hanging up, you remind her that you both agreed beforehand to skip Day One. This doesn't seem like a good idea because it means you're not going to give her your first impressions of the trip until Day Two. But you think, we need to stick to the plan, so you say, "I'm probably going to just take tomorrow and catch up, get on their time zone, you know?"

This doesn't seem to satisfy your wife even though she tries hard not to vocalize it from five thousand miles away.

"Uh, okay. We're pretty busy tomorrow anyway."

You hang up exhausted from jet lag, yet content that your first phone call only cost about US $4.

On Day One, you want to call but you can't break the agreement that early, can you? No, that would be a sign of weakness. Sure, your wife and kids have your number so let them call if they must. Let them be the weak link. But of course they don't because they are out to prove they're as tough as you. Therefore, there is zero communication on Day One and it kills everybody, although nobody will admit it.

The Day Two call is the best of all the calls. Everything is just as you hoped it would be. Your children, anticipating your call, eagerly wait by the phone and pick it up on the first ring.

"Hi Dad…we're okay…I played softball and Amy played soccer…where are you?...what TIME is it there?...is it fun?...Okay Dad, we have to go. I'm going to a friend's house and Amy has to practice piano. Want to talk to Mom?"

Now that's a great call! Not only are your children coping with your absence, they're not really even sure you're gone. Whatever were you worried about?

Your children are your most important concern on Day Two. Your wife should be able to tough it out until Day Four. Sure, you talk with her but it's small, pleasant talk: "How was your day?… anything interesting in the mail?… wish you were here." It's all very cordial.

Then you make one slight mistake.

You introduce *cost* into the conversation.

"Well, this call's probably getting expensive so why don't we call it a night, or in my case a next day," you say.

You try to broach this subject in a joking manner but it doesn't work and the damage is done. Your wife says nothing but files it away about halfway back in her head where it's easily accessible during a later conversation.

You hang up. In spite of the cost faux pas, you feel so good that you don't even regret skipping Day Three, the most stress free day on your trip. You know the wife and kids are fine because you talked to them yesterday. And you will talk to them tomorrow. On Day Three your phone stays holstered because you've vowed to make Day Three YOUR day. From the moment you get up, it's all about YOU!

On Day Three, things just go your way. You always seem to have some unexpected free time. You pass a scenic European park while you're wearing your jogging shoes. You weren't planning to jog but you just can't resist. So you begin a leisurely jog but stop mid-run because you happen upon a commercial shoot for a French perfume. And this commercial stars two equally hot French models.

You exit the park feeling healthy in every way. Even better, you didn't get lost. You know exactly where you are and you continue on your journey armed with the French model story, one that you will be telling your neighbors for years.

You get lunch on the street and have the exact amount of foreign currency in your pocket to pay for it. At dinner you use your company credit card and order whatever looks good, since you aren't paying for it. You eat at a bar and it just so happens that the guy next to you is foreign yet speaks excellent English so you have an animated two hours of conversation, discussing topics that you'd never talk about at home, like why every kid in the United States plays in four soccer leagues, six days a week, yet as a nation, the US still kind of sucks at soccer.

So there you are, enjoying YOUR day, totally unaware that SOMETHING is happening at home on Day Three. That SOMETHING is never a good SOMETHING. It definitely involves at least two of these subjects:

1) Stitches
2) The transmission

3) A possible fracture
4) A totally unexplainable, out of the blue "F"
5) Your mother
6) The phrase, "I haven't had time to even THINK about dinner"

THAT'S what is going on in your house on Day Three.

Day Four…TIME TO CALL. Remember, you have no idea what went on at home during Day Three. No, you're still feeling great from that massage you had on the same day. So you call. Your oldest answers the phone, the one who, for some reason, is suddenly ticked off at you.

"Hello?"

"Hi honey, it's Dad."

"Hi."

"What are you doing?"

"Getting ready for school."

"So you're probably rushing."

"Yeah."

"You being nice to your sister?"

"Mmmm." Then, "Want to talk to Amy?"

"Uh, sure. Have a great day at school."

"Bye."

Indians have had more meaningful conversations using smoke signals.

Now your second child, the youngest, gets on the phone and instantly makes you want to hail a taxi for the airport.

"Hi Daddy. You sound far away."

"Well I am honey. Remember when we looked at the map and I showed you…"

"When are you coming home?"

You crank the volume on your rented phone but it's no use. It's not the connection; you realize your daughter is talking in a whisper, while trying to stifle sobs.

"Not for another four days, princess. We talked about that too, remember? But Daddy's already been gone four days. In four MORE days I'll be home. That's not that long, right?

"It seems like a long time."

"I'll be home before you know it. I miss your hugs and kisses. Can I talk to Mom?"

"Okay. I wish you didn't have to go away. Ever. Ever ever again. I'll get Mom."

In the waiting silence that follows, you realize you are zero for two. One child hates you and one thinks you are orbiting the earth in the space shuttle. Then your wife picks up the phone. Her greeting is not warm and fuzzy but direct, as if your child had handed her the phone and said, "There's a man on the phone who wants to know if you'd be willing to take part in a brief survey."

HER: Hello?

YOU: Hi, honey.

HER: (SLIGHT UNCOMFORTABLE PAUSE) What are you doing?

YOU: Uh, talking to you. What are you doing?

HER: A little of everything. Actually a lot of everything.

YOU: I missed talking to you yesterday. What did you do then? (Remember, yesterday was Day Three. YOUR day.)

HER: (LONG SIGH) Well, Natalie had gymnastics but had to leave early. She said her foot is hurting. Her coach said something might be fractured (#3). We drove home. By the way, the car doesn't sound good (#2). Then I looked at her homework. Do you know she got an F (#4) for not turning in an assignment?

(**Note:** *Day Four is the day your wife forgets you have been gone for four days and therefore would have no idea about the "F."*)

Anyway, we didn't get home until eight. I was just trying to get the kids to bed when your mother called (#5)…What did you do yesterday?

YOU: (THINKING QUICKLY, KNOWING YOU HAVE TO LIE.) Nothing much. I'm still pretty tired from the flight.

HER: Well, at least you're by yourself. I haven't even thought about dinner tonight (#6).

YOU: Yeah, um okay. So what happened with the missed assignment?

HER: It's a long story. It would be too expensive to talk about now.

(BOOM! The cost factor just leaped from the middle of her forehead, catapulted directly through the phone line and lodged quite painfully in your ear. How could you have been so stupid? But as you are mentally slapping your brain with your fist, she rescues you from having to continue the conversation.)

HER: I'll tell you about it when you're home.

YOU: That's only four days from now.

HER: Uh huh. Seems like it's going fast, I guess. Do you think it's going fast?

YOU: Yeah. I suppose.

HER: Okay, I've gotta run. When will you be around tomorrow? Can I call you?

YOU: You mean the next day? Tomorrow would be every day, not every other day.

HER: (LARGE, DRAWN OUT SIGH) Okay, whatever. Call me.

YOU: I'll do that. Love you.

HER: I love you. Bye.

The "bye" is what you remember; not the "I love you."

Day Five is the hardest day to stick to the "call every other day" calling plan. Day Four's call went so horribly that you want another chance. You don't call but you have a miserable day anyway because every sight, sound, and decision comes with guilty overtones. You don't stroll the cobblestone streets in the evening, poking your head into assorted pubs and engaging in conversation. Those activities came and went in Day Three. Instead, you eat dinner in the hotel's restaurant, without attempting conversation. Whatever you ordered tastes bad and the wait staff seems so indifferent to your mere presence that, if a flaky European croissant lodged in your throat, you would have to perform your own Heimlich maneuver.

Day Five is the first day you turn on European TV in your room (or in my case, my cruise "stateroom").

You discover quickly that it is the same as American TV in that it has happy-looking (for Europe) news anchors, a home shopping network selling European crap, sports that you don't care about and a few American movies (dubbed in whatever language they speak in this country) that you recognize but never bothered to watch at home.

It differs only in that at least one channel—and possibly more depending on the time of day and your location—is showing porn.

Free porn.

Free *hardcore* porn.

The channels come up randomly as you click the remote; there is no rhyme or reason why the couple having sex is sandwiched (for lack of a better phrase) between CNN and *futbol* highlights. Or why another couple, this time both female, occupy the channel right past the cooking show. You tell yourself you don't feel like watching but you can't stop. Two hours go by and you're still awake.

You wake up groggy on Day Six and decide you are ready to go home. Unfortunately, you have two days left and it's brutal. Everything you see reminds you of home. The thousand-year-old castle perched high on the bluff looks inviting but not as homey as your back patio. You're drinking heavy German beer from a massive stein but it's not Bud Light (and never will be if you're German). You miss your family, the life you know and the comforts that go with it. You can't wait to make that phone call just so you can hear those chipper, sweet-sounding voices (from Day Two only) that have crept into your head and refuse to leave.

Day Six is the day your long-awaited call home kicks to voice mail.

MACHINE: Hi, we can't make it to the phone now. Leave your name and number and we'll call you back.

YOU: Hi, it's Dad. Was hoping to catch you guys. You have my number so call me when you get a chance. I miss you.

You hang up, convinced your loving family has decided you are taking up permanent residence in Europe and has thus, moved on.

The only thing comforting about this is realizing that, if you ever get a terminal disease, at least you know your family can exist without you. Between tears, you will say, "It's gonna be okay. Daddy will be in heaven. He won't be with you but that's not so bad, is it? Remember when Daddy went to Europe?"

Eventually your rented phone rings but you're the only one that seems to have time to talk. Your family has scheduled the phone call right between assorted practices, car pools and dinner on the run because, well, that's the only time they were all actually home together.

Like Day Four, the conversation is short. But at least there is a hint of anticipation in everyone's tone. Your oldest no longer seems to despise you and your youngest is less pouty but still pouty enough that you buy both kids another present each to stick in your carry-on luggage.

Whatever crisis occurred on Day Three seems a thing of the past. Your wife never brings it up during the Day Six conversation. She tells you about plans she's made for the next few days, plans that sweetly include the phrase, "if you're not too tired." Even though you are coming home in two days, you no longer care that this call costs eighty cents a minute. It doesn't even bother you that your youngest puts the phone down to find Mom and you are on hold for at least five minutes.

(On a side note, I've always wondered why our house seems to quadruple in size when I am away. Our house is two stories and four thousand square feet but I feel there must be secret passages, tunnels and hidden rooms that I don't know about because, when I ask one of our kids if they can "get Mom," they do just that and then I am waiting for an eternity before Mom is actually found. In the meantime, I'm treated to muffled sounds of "mom....MOM... MOOOOOOOMMMMMM..." over the phone.)

Day Seven is the day you break the rule.

You weren't planning to but it couldn't hurt, right? You've already packed, taking extra care with the gifts you purchased abroad. You head out for one final European meal and you see something along the way that makes you reach for the phone. This time, your wife answers.

YOU: Hey!

HER: (SURPRISED) Hey! I didn't expect to hear from you. This is the off day, right?

YOU: I know but I was walking down a street and saw this little café and thought about how nice it would be to sit there and sip wine with you. Next time, you're coming with me.

HER: That sounds so nice. I'd LOVE to be there now.

YOU: No more eight day trips. I PROMISE.

HER: It wasn't so bad. And besides, it's part of your job, honey.

Again, your spouse says just the right thing at just the right time. How sweet. Not only are you forgiven for anything that may have occurred while you were gone (not that you could control anything that did in fact occur at home while you were on another continent), but if another eight-day business trip should ever arise, you might be going again. Call the masseuse!

You keep talking. You don't care what the call is costing or that you will be home in twenty-four hours and could easily have this conversation face-to-face. For free. Tomorrow, when you open the door to your house, you want to make sure that EVERYTHING is totally cool and that you are up to speed on the events in everyone's lives. Even if your last business trip was a three-year tour of duty in Iraq, it's still awkward, when your spouse says, "Are you planning to come to school next Friday?" to respond, "What's next Friday?"

So you talk. More than forty-five minutes goes by and you are still talking. Remember, this was the day you weren't supposed to call but who cares? The Day Seven call is horrifically expensive; you didn't stick to the plan and, when you hang up, you feel like someone who returned to a three pack a day nicotine habit one second after Lent ended.

But you take solace in the fact that the Day Seven call will be the last you make with your rented, international, eighty-cents-a-minute, needs-to-be-mailed-back-IMMEDIATELY-or-God-only-know-what-you-will-be-charged phone. Your next call occurs on Day Eight. You make it in your country, with your phone and it doesn't matter who picks up the other end when you dial. It's the shortest call you've made in over a week.

It starts with two words.

"Daddy's home."

CHAPTER 10

Take This Job and Love It

A loud, dull thud and a squealing of brakes jolted me from my morning slumber. The alarm clock read 5:33 a.m.

I figured this disturbance could be one of two things: our quiet suburban hamlet had just experienced its first drive-by shooting, or the morning paper had arrived.

I ambled downstairs and outside to find that, thankfully, the latter was true. With coffee in one hand, I watched as our paperboy woke up every neighbor on the block with his paper-delivery technique. Actually, calling him a "boy" is like calling Madonna a "lady." In reality, he is a twenty-seven-year-old man who drives a Pinto in need of a muffler and less powerful stereo speakers, smokes cigarettes in the pre-dawn hours, and can heave papers out of both windows of a moving car while somehow keeping it from jumping curbs.

It's also very sad, I thought. What is this world coming to when the *Chicago Tribune*, self-dubbed "the world's greatest newspaper," is forced to hire fully grown adults to deliver papers? That's a kid's job and should always be a kid's job. Unfortunately, try telling that to today's kids. Today's kids are, as we speak, being hustled from soccer to volleyball to piano to taekwondo to extreme motocross lessons by their haggard parents. Kids with no athletic ability are sitting at their home PCs, making boatloads of money helping established companies with their computer problems. Several years ago, I read that Mattel employed a group of children to monitor and critique its Barbie doll website.

What? No direct link to Barbie toothpaste? I haven't seen anything that dumb in all my eight years!

In effect, these kids are the bosses of the company. Let's hope this trend doesn't continue, or tomorrow's workforce is going to find itself answering to two-year-olds.

SECRETARY: *Good morning, Master Billy.*

BILLY: *What's good about it? I didn't sleep at all last night. My new teeth are killing me. NOW LISTEN UP EVERYBODY! Look alive today. We've got a lot to accomplish before my nap. There's a shareholders meeting in twenty minutes and I want to make a good impression even though I don't even know what that means! So let's get cracking. Miss Johnson, we need snacks in the boardroom.*

SECRETARY: *I've ordered the Cheerios and juice boxes, sir.*

BILLY: *That's why I pay you the big bucks, dollface. Also, see that a copy of the 10K report is on my bouncy seat. Finally, somebody needs to wipe me ASAP. LET'S MOVE PEOPLE!*

During the late 1990s, also known as "the time when everybody thought they were stock market experts," newspapers were full of stories about twenty-year-olds who invented innocuous computer programs in their garages and then sold them to Fortune 500 companies for millions of dollars. That trend subsided somewhat as the new millennium dawned, dot-coms folded, the housing market tanked, unemployment skyrocketed and a concerned President George W. Bush watched it all from a box seat at the Washington Nationals game. Yet there were still "success" stories like Mark Zuckerberg, who created Facebook from his Harvard dorm room. By 2009, the now twenty-five-year-old Zuckerberg was number 158 on the Forbes 400 list of the richest people in the United States, with a net worth of $2 billion. Donald Trump's net worth was also listed at $2 billion in 2009, but he was ranked ahead of Zuckerberg thanks to some quick phone calls from the cast of *Celebrity Apprentice* along with a massive Facebook campaign.

I'm sorry but when you're twenty-five, your idea of splurging should be ordering the LARGE fries. You should be cruising for women in a beat-up Camaro, not your own jet or Hummer. You should be happy sleeping on a friend's pull-out couch, not the penthouse suite at the Four Seasons.

So as I write this today, I admit that I am hatching a crime. If you feel my diabolical deed needs to be stopped, feel free to call the Illinois State Police, the FBI or your local juvenile authority. Otherwise, stand back and let me enjoy my foray into the world of torture. Maybe Anthony Hopkins will play me in the movie.

Okay, here goes. I plan to kidnap three high school males and force them to work at one of my high school jobs for one summer. I promise not to physically harm any of them although they will probably insist they are being treated worse than Central American children who work in shoe factories. My three victims will not make nearly as much money as Zuckerberg but they will learn what it's like to deal with people as opposed to technology. They will learn the value of self-discipline. Maybe they'll even have a few laughs along the way. Oh, and they will also get great tans.

After reading that last sentence, you've probably guessed that none will be working in a fast-food restaurant. I was somewhat of an outcast in high school because, as a teenager in the late 1970s, it was almost REQUIRED that you spend a portion of your free hours wearing a paper hat and smelling like deep fried lard as a result of your employment at McDonald's, Burger King, Wendy's, Kentucky Fried Chicken, Taco Bell or any one of a dozen independent grease traps that littered the landscape in my hometown. Even though I wasn't a fast-food employee, I had friends who were. This had its advantages. My buddy Phil worked at Taco Bell and, at least once a week, we cruised over there during lunch hour and stuffed our faces with Burrito Supremes before returning to class. When you're sixteen, the projected pitfalls of eating burritos and then doing ANYTHING never occurs to you. All we cared about was that the food was FREE!

Okay, back to my kidnap plot. I'm combing the high schools now in my wife's mini-van. It's passing period so there are plenty of ripe youth outside doing what typical high-school kids do: swilling

Mountain Dew, texting five friends at once, and drowning out the outside world courtesy of their iPods.

Ah, there he is, my first victim.

Hey kid, get over here. What's your name? Jeremy? How old are you? Fourteen? Well Jeremy, get in the car. Just do as I say or I'll force you to drink the contents of the Sippee Cup, which I just found behind the car seat and looks, judging from its contents, to have been there for about four years. Uh Jeremy, do me a favor and pull up your pants. Pull them up further. I can STILL see your crack. I don't want to see your crack and neither does your chemistry teacher. Actually, take your pants off now and put on something more "conservative." No, I'm not some kind of pedophile. But those pants aren't gonna fly when you start your job. That's right Jeremy, you work for me now. And beginning tomorrow, you're going to deliver papers, just like I did when I was your age. Better get some sleep.

I first entered the employment ranks as a paperboy for the *Daily Herald*, a surprisingly decent local newspaper in that it actually reported news. Most local suburban papers lead with stories like, "SIX-PACK STOLEN FROM GARAGE. VETERAN DETECTIVE SUSPECTS THIRSTY KIDS!"

A case in point was the *Garland Daily News*, a suburban Dallas paper that offered me a job shortly after I graduated from Northwestern. There was an opening in the sports department so I flew to Garland for an interview, envisioning myself as the hotshot cub reporter for the Dallas Cowboys.

That vision lasted about thirty seconds.

The sports page at the *Garland Daily News* was just that—*a single page*. The other *six* pages of the entire paper were devoted to more pressing matters, such as the public school lunch menu. Did an eight-year-old's culinary choices really constitute news?

Oh my God, Wednesday is tuna casserole. Lock the doors!

Trust me, if a six-pack had been stolen from a garage in Garland, the *Daily News* would have produced a five-part series and then submitted it to the Pulitzer committee.

Hey, there's an alcohol-related crime problem in the Heartland and we've exposed it!

The sports editor introduced himself with the standard Texas greeting:

"HOWDY GREG! WELCOME TO THE BIG D!"

I have always hated Dallas for that reason. In what other city do the natives refer to their home with a single initial? Chicago is not called the "Big C." Salt Lake City residents do not greet outsiders with, "Welcome to the midsized-but-growing-rapidly SLC!" What made it worse is that Garland was not, in any way, "the Big D." It was the "dusty, tumbleweed-infested, pick-up-truck-drivin' G."

The editor told me I would be spending most of my time covering American Legion baseball. I wondered if that meant covering the game itself or the postgame party at the sponsoring bar. Either way, this did not seem like a career stepping-stone so I waved goodbye to Garland and its tumbleweeds.

I returned home and applied for a reporting job at the *Daily Herald*, which politely rejected me. Apparently my service as a paperboy carried no weight. The *Daily Herald* already employed a hefty staff, had veteran "beat" reporters and occasionally scooped the *Chicago Tribune* and *Chicago Sun-Times*. Heck, it even had a

Thursday food section, crammed so full of food-related news that there wasn't even space for school lunch menus.

Okay, Jeremy are you awake? You should be. It's 5:00 a.m. Now open your front door. You should see a stack of newspapers on your step. They'll be bound with a piece of wire. You'll have to cut it with a knife. Got that done? Good. Now sit your butt down and roll every paper. That's right, I said roll it! Now bind it with a rubber band.

Seriously, when was the last time you grasped a paper held together by a rubber band? The guy with the Pinto, if he feels like it, puts my paper in a plastic bag, which he then fires into the nearest puddle in the general vicinity of my driveway.

Hey Jeremy, you're not tired yet, are you? Remember, we still have to deliver these papers. Hey, where are you going? No, Daddy and his Lexus aren't going to help you. You're going to do it yourself. Quit eyeing your collapsible scooter. You won't need that. Instead, you're going to WALK!

My route was fairly small by newspaper standards; about forty-five papers over a three-block area. Since the geographical terrain was so compact, I chose to walk when delivering papers. I loved seeing my neighborhood awaken, loved the smell of the early-morning dew and the sight of a fellow early riser walking his dog.

Okay, I lied. The real reason I walked is that I had just purchased a Schwinn Varsity ten-speed bicycle and the idea of a wire newspaper basket clipped onto the front handlebars would just add to my exterior dweeb appearance. Why make it worse?

While walking, I made a game out of the "paper toss," counting how many front steps I could hit from the sidewalk. Errant throws usually crashed into front screen doors so loudly that slumbering dogs began barking. No matter how inaccurate my aim was on a particular day, most customers were already up and awaiting my arrival. Many were elderly and considered the paper the highlight of their morning. The news, sports, obituaries and crossword puzzle kept them occupied until their afternoon highlight: getting the mail. Some customers actually set their watches to my arrival. If I was late, they were at the door saying, "I was about to call the paper and ask where you were." As I turned to leave, I wondered what they would do if I didn't show up at all? What would their days be like without news? Without weather? Without recipes? As strange as it sounds, delivering papers filled me with a sense of warmth—a sense that I was making a difference in these people's lives. I'm certain my friend Phil did not share this feeling as he stuffed refried beans into tortillas. Likewise for today's kids who design websites from their homes. It's hard to feel warm and fuzzy when you're contemplating using your salary to purchase an Escalade.

Hi Jeremy! Home already? That wasn't so bad, was it? Did you stop to talk to anybody? Tell them how high school is going? People aren't so bad, are they? What? Where is your money? Oh, I almost forgot. You have to collect the money yourself. Better hurry. The weekend is approaching and you want to be sure you have some cash so you and your buddies can go to the multiplex and see Confessions of a Sorority Strangler.

Instead of paying the *Daily Herald* directly, my customers paid me. I sent that money to the paper, which then cut me a salary check twice a month. Okay, it seemed like a strange way to do the books but who was I to question accounting methods? All I knew is that this was how I was going to make money. Therefore, it was with great pride that I ventured out every other Saturday, armed with a sheaf of red cards looped on a metal ring. I felt like a fourteen-year-old Tony Soprano, coming to shake down my clients for money. The only difference was that I could not threaten them with bodily harm if they chose not to pay. What was I going to do? Bite them with my braces? Melt them with my eyeglass lenses? Luckily most were eager to pay, even throwing in some tip money around the holidays. I recorded every transaction with a hole-punch, deftly piercing a number on the red card, a signal that they remained in my good graces. As I remember, the weekly paper fee was $1.60 and customers almost always paid in cash. Somehow I doubt the twenty-year-old software inventors use this method when selling their multimillion dollar products.

What's up Jeremy? Back from collecting? Your wallet looks pretty full. You must have made some tips. That means you were actually recognized and rewarded for doing a good job. Now go to your room and count the money. Put a little of it aside for the new car you'll have your eye on in two years. Put the rest in your pocket. Have fun at the multiplex. Make sure your friends see the contents of your bulging wallet. Tell them how it got that way. Maybe it will give them some ideas. You've only got two more months and then you're released. Nice job. You won't be seeing me anymore. I'm headed back to your school to pick up my next victim.

I move the van slowly through the high school parking lot. It's 2:30 p.m. and classes are just letting out. Shouldn't be too tough to find a kid with a severe slouch and attitude to match. Wait, I see him.

Hey kid, got the time? Come closer so I can hear you. Come even closer so I can get a look at your face. You see, my vision is blinded by that streak of purple in your hair. And those four studs in your ear, do they represent anything? You're not in a gang, are you? It's just that I noticed that tattoo of a skull mounting a devil and I wondered. Why do you have a tattoo anyway? Do you know what that's going to look like when you're sixty? How old are you now kid? Fifteen? What's your name? Connor? Well Connor, GET IN THE VAN. Don't say anything, you're going to be fine. We're going for a little drive. First, we're going to have that tattoo removed. You also won't be needing your cell phone. Why? Because the members don't want to see you texting anybody. That's right, I said, "the members." Welcome to your summer job as golf caddy.

When I was fifteen, my office was Rolling Green Country Club in Arlington Heights, Illinois. The name itself implies a certain degree of snobbery and the club certainly delivered once you passed through the iron gates. At Rolling Green, the fairways were both rolling and green, the sand traps perfectly raked, the valets well groomed and the roast beef medium but never too rare.

Alas, the golfers still sucked.

Of course, I didn't find this out right away. I wasn't sure if I wanted to caddy at all. The paper route was still providing me with spending money and the idea of lugging a golf bag around in 90-degree heat seemed about as appealing as cutting the front nine with nose hair trimmers. Plus, I still hadn't entered puberty at fifteen and most of the golf bags were taller than me. The average country club golfer carries about forty clubs. He brings his own set while "trying out" another set recommended by the club pro who, in an attempt to say something positive about the member's game, tells him these other clubs might give him a bit more "pop." Then, for good measure, he throws in a few more clubs that he picked up on his last trip to Scotland. Add four dozen balls, a rain slicker, two pairs of shoes and an umbrella that would keep the entire state of Rhode Island dry and you're talking about 120 pounds.

In spite of my trepidation, I decided to give it a try. It paid better than paperboy and would occupy more of my summer, much to my parent's relief.

Hey Connor, quit whining. Do you play golf? You do? That surprises me. Doesn't your nose ring interfere with your vision? Well, if you work hard at this job, you can play the course for FREE on Mondays. The members pay $50,000 in initiation fees to play it. Oh, and are you thinking about college? Ever heard of the Evans Scholars? These kids go to college for FREE just because they caddied in high school. Now put on a clean shirt and lose the attitude. You're about to meet the caddy master.

Just as private country clubs today are struggling institutions, caddy master is a dying profession. But in the 1970s, you can bet that every private club with a golf course had one. The master's sole responsibility was to train the caddies in the spring and make life miserable for them all summer. On a warm afternoon in May 1977, I met Gil, the Rolling Green caddy master. Gil was about fifty, an ex-army man with an amazing ability to smoke a cigarette until it disappeared entirely. While other nicotine lovers disposed of their completed smokes by stubbing out the butts, Gil merely rolled his thumb and forefinger together. Whatever was left of the cigarette floated harmlessly into the grass, becoming fertilizer. Gil's yellowing fingernails and nicotine-scarred voice were proof of his habit and his skill. His vocabulary also would have made most drill sergeants go running into a confessional with their ears covered.

Bless me, Father, for I have sinned. I have just heard the Lord's name taken in vain so many times in a single sentence that I fear I am going to secondhand Hell!

One day Gil emerged from behind the "starter's cage" to announce that the toilet paper kept disappearing from the bathroom and from now on, "you little bastards are going to have to use your goddamned fingers."

Gil's caddy school consisted of telling his new charges how to hold the bag, where to stand during shots, how to rake a sand trap and most importantly, how to "stay out of the fucking way." After about thirty minutes of his abuse, I was officially a caddy.

My first "loop" occurred that weekend, when my badge number was pulled from the caddy bucket. All caddies wore badges, adorned with their names and a number. The name was so the members knew what to call you. The number was for Gil's benefit; he didn't believe caddies had earned the right to be referred to by name.

When the caddy shack opened each day at 6:00 a.m., all badges were deposited into a white, plastic bucket and pulled by Gil, who screamed the numbers in between puffs. Once. Only once. Toilet paper or no toilet paper, all caddies knew not to use the bathroom between 5:55 and 6:15 a.m.

The process was agonizingly tedious; it was like waiting to hear your name called at the Department of Motor Vehicles. The badge pull determined the order in which caddies were assigned "loops." If Gil pulled your badge early, you could be home by noon with money in your pocket. Even better, you could hang around and hope for a "double loop" later that afternoon. But if you were farther down the list, you risked having to suffer all day watching a broken television or playing pool on a table that sloped so badly, Olympic ski jumpers trained on it.

I breathed a great sigh of relief when my first loop turned out to be female. Even though I had never caddied before, I knew I would enjoy caddying for women more than men. For starters, their bags were lighter. Unlike their husbands, they didn't feel it necessary to carry every piece of golf equipment ever invented. They just wanted a nice, light bag that matched their outfits. Sure, they carried all the clubs in a standard set but they only used three of them. Typical club selection for a country club woman golfer went like this: driver off the tee, followed by one club which she kept hitting until she reached the green. It didn't matter whether she was fifty or 350 yards away; she had one club that she hit a great shot with about two years ago and had never forgotten it. Therefore, she kept hitting that club—usually a 5- or 7-iron—in hopes of recreating the magic. Finally, upon reaching the green, she whipped out a putter that was usually pink with a daisy decal on the blade.

What I liked most about women golfers is that they DIDN'T CARE WHAT THEY SHOT. They took up the game only after marrying their husbands and realizing golf was the only way to spend time together. Notice that I didn't say "quality" time. Just time. When women played together, they talked about mall sales, beauticians and their favorite salons. They gossiped about other club members. And they did it all while in the middle of their backswings. They hit the ball 150 yards maximum and always hit it straight, making the caddy's job immensely easier. Hit it in the woods? No problem. They simply dropped another ball in the fairway, swatted at it with a seven iron and continued their conversations.

The male members were a different story, as my victim Connor is about to find out.

Hey Connor, I just heard your number called. Ready to go? You look a little scared. Oh, the caddy master just chewed your ass because you were wearing your badge on the right side of your chest instead of the left. You'll learn soon enough. Is your towel wet? Then get out there because they just brought your bag to the first tee. See it? It's the brown leather, two-story model. All the veteran caddies recognize that bag. It belongs to Arthur Fontaleoni, the biggest jerk at the club. This guy founded a chain of plumbing supply stores and sold the business four years ago for $250 million. He plays here every day, usually with the same three guys. Rumor has it they play for as much as a grand a hole. Let's hope Fontaleoni brought his "A" game today. He hates to lose.

Here comes Fontaleoni now. Did I just hear him refer to you as "son?" Get used to it. He's not going to read your badge. If he's playing well he might converse with you about the ninth hole but until then you're going to be "son" or "kid." You've got his driver out, that's good. He's ready to tee off. Uh oh, he just duck-hooked one right into the woods. You saw it, right Connor? Fontaleoni hates to lose a ball. If he hits one into a rock pit full of venomous cobras, he'll still look for it. No, he won't; he'll send you to look for it. And if you can't find it, he's liable to club you over the head with a pipe wrench. The betting has begun. Have fun!

As if paying thousands of dollars to belong to a private club wasn't enough, Rolling Green's male population had to make it even more interesting by betting on every shot. Don't country club members know that golf course gambling almost destroyed Michael Jordan's career? The members spoke in a language I didn't understand but I knew that if I heard the words "Nassau," "press," "sandies," "greenies," "skins" or "carryovers" on the first tee, then I was in for a long afternoon. To me, nothing ruins a relaxing game of golf quite like betting. Mild-mannered men become complete idiots when a missed three-footer costs them $150. Also, because a round of golf typically takes about four hours, there is plenty of time to add up your losses in your head and become even angrier. And nobody feels the wrath of an angry golfer more than his caddy, whose status is just higher than that of antebellum slave.

How's it going, Connor? What's Mr. Fontaleoni doing now? He's got a ten-footer for birdie? I hope he makes it. Too bad, he missed it. Uh oh, he missed the one coming back too. Ouch, three putts from ten feet. No doubt, that cost him a few g's. Wait, what was that piece of metal that just went flying by? His putter? He threw his putter into the next fairway? Well don't just stand there, Connor. Go get it. And hurry up because Fontaleoni's standing on the next tee, wondering where the hell you are.

I'm not saying all Rolling Green golfers were this horrible. When I caddied, some were downright pleasant. They asked me about my interests, what school I attended and what I planned to major in once I started college. Some even asked me for advice on club selection or wanted me to read their putts for them. The most valuable lesson I learned from caddying was how NOT to treat people. While caddying I vowed that, if I ever got to employ a caddy's services, I would want him to enjoy the caddying experience. My wish came true about fifteen years later at, of all places, Rolling Green Country Club. You may not believe this but my parents decided to JOIN the club. Apparently my rants and raves as a teenager about the horrible members fell on deaf ears.

I joined my Dad for a round shortly after graduating from college and enthusiastically agreed to his offer of a caddy. When I arrived at the first tee, Doug was waiting for me. I knew his name because it was on his badge—the same style that I wore eight years before. Doug looked nervous. Or maybe that look was him sizing me up and wondering if I was going to be a nice guy or another Arthur Fontaleoni. I set him at ease immediately.

Hey Doug. I'm Greg. Call me Greg. I'll try not to hit any balls in the woods but if I do, we'll look for them together, okay? Do you play golf, Doug? Yeah? Are you on the team at school? How's the team this year? How long have you been caddying? Do you like this job? People can be jerks some times, can't they? Don't worry; you can be honest with me. You know, I used to caddy. Let me tell you about the moron I caddied for one time…

Well Connor, you've reached the eighteenth green. You look beat. After all, the heat index today is 103. What did Fontaleoni shoot? 98? Wow, you could probably cook an egg on his forehead. Any idea how much he lost? $2500? Ouch! Don't expect a big tip. He needs everything *he's got left for the gin rummy game in the locker room. What have you learned today? Hopefully you learned how having a lot of money doesn't necessarily make you a nice person. And hopefully you've learned how much people appreciate a friendly comment here and there. You've learned the world won't come to a screeching halt if you and your cell phone are apart for a few hours. Finally, you've learned a little bit about manual labor. Next time you see the janitor in your high school fixing a leak or repairing a piece of drywall, don't turn to your friends and make a snide comment. Give him a smile and ask him how his day is going. Now get back to the caddyshack because there's an afternoon shotgun starting in thirty minutes and they need caddies. Can't you hear the caddy master now?*

GET YOUR BADGES IN THE BUCKET, YOU LOW-LIFE FUCKERS!

Well, I'm back in the minivan. It's been a good day so far. But I'm saving the best for last. My next victim has no idea what fate awaits him. I might have to look somewhere other than the high school. There's a Best Buy up ahead. I'll just park my van and venture inside. There's some kid standing in the video game aisle, playing online *Mortal Power Kombat Ranger Navy Seal Death Squad.*

Hey kid, how do you play this game? Oh, you're trying to kill as many evil rebels as possible? What happens if you accidentally kill innocent, unarmed women and children? Oh, you lose fifty points, eh? Bummer. You're pretty good at it, kid. What's your name? Kyle? How many hours do you spend here Kyle? Until store security kicks you out, eh? Do you have a job? Oh, you sell a little weed here and there? Charming. Hey Kyle, your eyes are getting pretty wide. Is it because you feel something cold and hard sticking into your flesh?

Don't scream. Just follow me out to the parking lot and get in the van. NOW.

Wow Kyle, you are gullible. Did you really think I had a gun? That was my cell phone. You must be high. If I were you, I'd come down off Planet Teenager immediately because you're about to start your summer job. You're going to be a door-to-door salesman for the Fuller Brush Company.

Headquartered in Great Bend, Kansas, Fuller Brush has been around for nearly a century. Red Skelton even starred in a movie titled *The Fuller Brush Man*. The business plan was simple: employ an army of door-to-door salespeople, and send them out to middle-class suburban neighborhoods across America. Tell them to do the following: confidently march up to doors, engage stay-at-home housewives in friendly conversation and then present them with catalogues full of products like liquid degreaser, stainless steel scouring pads, berry-scented air fresheners, chemically treated dust mops and, of course, the signature line of hairbrushes. Fuller Brush products were known for their durability; as a Fuller Brush man I heard countless stories of hairbrushes that had survived tornadoes, hurricanes, wars, fires, dog teeth, divorces and other calamities.

I have yet to see a Fuller Brush person in my neighborhood but the company still has an established Web presence, and, according to an e-mail I received, a door-to-door contingent as well. However, the dwindling presence of actual salespeople may also have something to do with the fact that most door-to-door salesmen today would be greeted with a can of mace and a Rottweiler's jaws, even if they were selling winning lottery tickets. But Fuller

Brush salesmen thrived in the 1950s and '60s. My dad sold Fuller Brush when I was born; sixteen years later, he suggested I give it a try. At sixteen, I was becoming fairly efficient at the gift of gab (read: I had a big mouth) and Dad thought I might be able to pioneer my verbal talents into some hefty spending money. With a little research we discovered that Fuller Brush was indeed still thriving in Chicago's northwest suburbs. I made a phone call to headquarters, told them I was sixteen, endured a lengthy pause at the other end of the line and was informed that a regional supervisor would be in touch with me.

That supervisor turned out to be Ken, a geeky sort with slick-backed hair, horn-rimmed glasses and a pencil-thin mustache. Ken arrived at my house one Saturday and announced that he would be training me. Ken presented me with the tool of the trade—an actual leather briefcase loaded with some of the more popular Fuller Brush products including hand lotion ("let the ladies try it, they'll LOVE it!"), air freshener ("spray it in their house, they'll LOVE it!"), and the number one seller at the time, a carpet sweeper ("throw some lawn clippings on their carpet and vacuum them up. They'll LOVE it!").

After going over the product line for a few minutes, Ken and I jumped in his car and ventured into a real neighborhood, where I made my first sales call. Before sending me up to the first house, Ken gave me a stack of plastic spoons, perforated with tiny holes. He called them "tea strainers." I had never before seen a tea strainer in my life.

"These are the customer's free gift," Ken said. "When they answer the door, offer them one. It breaks the ice. And they'll LOVE it!"

I nodded obediently. "Free gift breaks the ice," I mumbled to myself as I approached my first screen door and knocked. A middle-aged woman wearing a hairnet answered.

"Fuller Brush," I said, reaching into my back pocket for a tea strainer and sending all twenty-five of them tumbling onto the front porch. "I have a free gift for you." (Just choose from any of these that are currently scattered all over your property.)

To my amazement, the woman accepted the catalogue and began thumbing through it. "Fuller Brush?" she cried. "Where have you guys been hiding? I LOVE your stuff."

Within minutes she had purchased a dust mop for ten dollars and I had earned four dollars. That's right, my commission was a whopping 40 percent! Sure, I felt like a bit of a dork carrying a briefcase through neighborhoods, silently praying that no high school chums would answer the doors. Nevertheless, 40 percent was nothing to sneeze at. My friend Phil certainly never made 40 percent of refried bean transactions.

Okay Kyle, put on a decent shirt, a pair of pants and some comfortable walking shoes. Here's your briefcase, your catalogues and your free gifts. I'll make it easy for you. We'll start in a fairly upscale neighborhood. Here's one. The lawns are well manicured and the homes are required to be at least three thousand square feet. Now all you have to do is walk up to the six-paneled oak doors and ask for "the lady of the house," just like I used to do. Don't roll your eyes, Kyle; I know it sounds stupid, but it also shows a little respect. Today people live behind peep-

holes drilled into doors that are wired to tiny security devices, which are hooked directly to central security stations, which have immediate lines to police departments, which are in direct contact with the Navy SEALs. If you come off the least bit suspicious, a SWAT team will be crawling up your ass in a matter of seconds.

But if you can portray yourself as the friendly, hard-working high school student that I know you can be, you'll be amazed at your success. Remember Kyle, people are generally friendly. They like seeing someone with a good work ethic. And today they love convenience more than ever. You're giving them a chance to shop WITHOUT LEAVING THEIR HOUSES! I know they do that online but you might alleviate loneliness for somebody. Amazon.com won't.

I sold Fuller Brush products for two years. In that time I expanded my territory. Basically, I went anywhere I wanted. I learned when to follow up with customers, how to anticipate their needs and how to read their body language as they perused the catalogue, a skill that often lead to additional revenue. Because I made most of my sales from inside people's homes, I saw how different people lived, how they treated their children, their pets and their houses. Most importantly, I learned the art of SELLING something, a skill they don't teach in high school and one you certainly don't get at today's fast food establishments or by sitting by yourself in front of a PC. To this day, I consider it to be the most valuable job I ever had.

Hey Kyle, how was your first day? You knocked on fifty doors? How many sales did you make? Five? 10 percent isn't bad for your first day. Some days you'll be shut out completely. What did your customers buy? A ladies brush? Good. Some floor finish? Excellent.

What's that you say? A lady wanted ten catalogues to distribute in the teacher's lounge at the high school where she works? Kyle, you may have just hit the jackpot. Before you know it, you'll have customers calling you. Your sales empire will expand before your eyes and then you'll be able to make some bucks just by picking up the phone. But until then, keep your briefcase handy because there's another subdivision full of potential customers awaiting you tomorrow. Now go fill those orders and send them to Great Bend, Kansas. The sooner they deliver the goods to you, the sooner you'll get paid. Enjoy your 40 percent.

Well, that about does it. Three kids, three jobs and, hopefully, three altered attitudes. Maybe I made a difference in one of these kids.

I can only hope.

CHAPTER 11

There's an App for that on Pandora

In October 2009, as the air turned cooler and the economic recession showed no signs of abating, I did what any logical budget-conscious parent would do: spent the Columbus holiday weekend in New York City. Some day history will show that Columbus chose not to actually *land* in New York because hotel managers, sensing his arrival, tripled their rates.

My companion for this trip was Natalie, then twelve. It would be her first time in the city, and probably the last time in the foreseeable future that she would want to be seen anywhere in public accompanied only by her dad. Promises to visit the Statue of Liberty, attend a Broadway show and stroll through Central Park on a sun-kissed fall day elicited the standard twelve-year-old response: "Mmm hmmmm."

Until she saw the Apple Store on Fifth Avenue. For a twelve-year-old, the Apple Store is the Emerald City.

Actually, calling it a "store" is like calling Treasury Secretary Timothy Geithner a "bank teller," as we discovered the moment we approached the massive glass cube bearing the ubiquitous Apple logo and descended the escalator into a single room. Personally I was expecting a multitude of rooms. But this was New York and even on Fifth Avenue, space apparently is at a premium.

So there we were surrounded by four walls, hundreds of iPhones, iPods, and iTouches, along with thousands of iGeeks. And everywhere I turned, I saw the catchy slogan that the Apple marketing team churned out for the iPhone: "There's an app for that!"

I'm sure Apple CEO Steve Jobs personally approved using "app" instead of "application" in iPhone advertising. It definitely sounds cooler and saves time in the event that iPhone users decide to converse verbally. However, after several years spent observing iPhone users tap, scroll, zoom, chop, slice, dice, wash, rinse and spin on their devices, verbal communication doesn't appear be a priority. The next iPhone slogan, currently undergoing testing, is, "Don't forget. You can talk on that!"

Over the years I have fattened Jobs' wallet by purchasing two iPods, countless iTunes songs, and even a movie which I downloaded to the device and tried to watch on a plane. I quickly gave up; Spiderman's webs just don't look as intimidating on a two-inch screen.

Yet, I'll admit, I think the iPod is the coolest invention since either the Post-It note or the flip flops with bottle openers in the soles. A company called Reef makes those, in case you're interested. I once dropped my iPod while jogging on a South Carolina beach and, upon realizing there was no app for removing sand from its inner being, gathered my wife and daughters together and solemnly told them there had been a death in the family. I imagined a five-mile run without those white buds stuck between my ears and the inability to listen to Neil Diamond, 50 Cent, Elton John, Indigo Girls, Counting Crows, Death Cab for Cutie and my favorite songs from *Les Miserables* back-to-back. I got so depressed I canceled my morning jog and consumed an entire carton of Ben and Jerry's Cookie Dough.

My daughter, however, would consider it a thrill to be adopted by Jobs's family, so into Apple technology is she. The iTouch was on her Christmas list for 2009, but she also fantasized about owning an iPhone and a MacBook, Apple's sleek laptop computer with the catchy slogan: "It does everything a PC is *supposed* to do." In April 2010, Apple also released the iPad and its equally catchy slogan, "It's either a smaller MacBook or a bigger iPhone!"

At the Fifth Avenue store, I watched her move from one device to the next, feeling slightly guilty for I knew that neither the iPhone nor the MacBook would be under the tree in two months. The MacBook was too expensive for the Schwem budget. The iPhone was eliminated for another reason: my daughter had recently *lost* her phone and I had no intention of reliving that experience with an iPhone.

When you're twelve, losing *any* phone—even one that must be cranked by hand—is worse than losing the lone key to your rich and newly deceased Aunt Gladys's safety deposit box. Gladys of course, is everybody's daffy aunt who slurps her coffee, smacks her lips and licks her fingers at the holiday dinner table. But in between these gross out episodes, she reminds everybody that, when she expires, she's got a nice nest egg in that box and everybody will be pleasantly surprised. Upon hearing that, relatives fall over themselves saying, "More turkey Aunt Gladys? And can I refill your coffee mug while you finish that cigarette?"

Natalie's Sprint Rant met an untimely death a mere nine months after she had sent approximately 1.3 billion text messages and 400,000 self portraits of her teeth and nostrils to her friends. Thankfully her plan included unlimited text. If not, the Schwems would be living in a cardboard box under a viaduct.

The phone failed to appear one morning as she was getting ready for school. This was no great shock as each morning at our household begins with a frantic search for shoes, backpacks, hairbrushes and homework. The item in question is always located approximately three seconds after the bus pulls away.

Sue and I began our detective work by asking the question burned into every parent's vernacular: "Where was the last place you left it?"

The answer was just as common: "I dunno."

Every time I hear that response, I wonder, what's the fuss over waterboarding?

Eventually we gleaned a little more information from her. She couldn't remember bringing it in last night. No wait, she remembered having it when she got out of the car. Okay, it's getting clearer. She remembered putting it down while she retrieved her backpack from our SUV's hatch.

And here's where she had a *CSI* moment—the case suddenly becoming crystal clear. She remembered placing it on the *bumper* of the SUV while she retrieved her backpack.

This would have been perfectly okay if the car had remained in the garage between 4:00 p.m. and 7:00 a.m. the next morning. But our family car is never idle for that long unless the battery is dead. No, our car had made at least three trips since 4:00 p.m., meaning the purple Sprint Rant had made at least one of them—without a seat belt, if you catch my drift.

Two days later, Sue found the phone. Check that, she found *pieces* of the phone along West 123rd Street. There were just enough parts and accompanying tire tracks to ensure it wouldn't be covered under the warranty's "drop" section.

My daughter took the news hard, knowing that a new phone would come from her allowance. But, as she began to save, she began looking at newer phones and cooler phones. And the iPhone is confidently nestled atop the "cool" chart. I have to hand it to tech guys; they have a knack for introducing newer, flashier and easier-to-use products mere seconds after a consumer has made a purchase. Microsoft recently unleashed something called Windows 7. I'm not sure what happened to Windows 1–6, but I am still trying to figure out how Windows 7 got released fourteen years *after* Windows 95. That's like Schick releasing its four-bladed Quattro and, twenty years later, saying, "Now we've even got one with *two* blades!"

Naturally, my daughter was fascinated by the apps. The last time I checked, the number of approved apps was well over 100,000, including dozens that center around—*ahem*—gaseous noises. I wondered if Jobs owned any of them. Specifically, had he downloaded *Crazy Fart Detector, Fart Button, Juicy Fart, Fart Match, Tip 'n Fart, Fart Keyboard, Fart Box, Fart for Free, Name that Fart, Fart Call, Fart or Die, Ms. Fart, Fart for a Buck, FartFX, Fart Shaker Deluxe, Fart Trivia, Fart Piano, Fart Lighter, Sexy Bikini Fart, Latina Bikini Fart* or *Hot Girls Who Fart?* That's right, all of these have been *approved.* God only knows what's in development.

I didn't nix the iPhone purchase right away. As I've stated in previous chapters, I'm an open-minded guy when it comes to technology, even if I don't always understand it. I firmly believe in trying things before forming opinions, something I have tried to do with everything except caviar just because it looks disgusting.

As dusk fell over the city, Natalie pondered a sudoku app while I grabbed a display iPhone and began browsing the online app store. What apps would I put on it should I choose to own one? And what would happen if *I* lost it? I decided to experiment.

I chose ten random apps, none of which involved farting. First I'd download *iOwn*, a $4.99 app that allows the inventorying of everything one has acquired over the years. It would be like having one of those metal self-storage sheds on my phone.

Okay, now that I have EVERYTHING I OWN on my iPhone, there is still room for nine more apps. So I'd add *Pennies*, a finance app that lets me keep track of my expenses; *Grocery iQ* for control of my shopping list; and *Barista*, so I could make my favorite espresso beverages with the skill of a tenured Starbucks employee.

I may be tempted to sweeten that frothy drink so my iPhone better contain *BloodSugar*, an app that allows me to test my sugar intake. I'll add *FlightTrack* since I spend a lot of time in the air and need to get REAL TIME FLIGHT UPDATES! *Gas Cubby* will record my gas mileage in the event that I miss my flight altogether. *RedLaser* lets me scan UPC codes and pay for stuff via my iPhone while *DIRECTV* allows me to program my DVR from faraway places.

Finally, I'll add *Things* which, according to the Apple website, allows one to "manage tasks and get things done." I assume nobody in Congress has ever downloaded it.

Now I will drive down West 123rd Street with my newly loaded iPhone on my car bumper until it falls off and becomes pothole filler. In one instant, I have suffered amnesia of every sort. I no longer have any idea what to buy at the grocery store and without a UPC scanner, wouldn't know how to buy it anyway. Of course, lacking *Pennies*, I'm not sure I can afford groceries period. Or gas for that matter.

Since purchasing coffee from a store is out of the question, I'll just go home and make some. Wait, no I won't because all the recipes were stored in *Barista*. And I probably couldn't find the coffee maker anyway since I downloaded its location on *iOwn*. My next flight came and went (I think) and my DVR is suddenly useless. So I will sit in my house, unsure what to do since *Things* is not around to guide me.

Wait, I'm feeling light-headed! Could my blood sugar be plunging? How should I know? Quick, find a phone and dial 911!

Damn, I can't do that either. Cause of death? Missing iPhone. Bag him and tag him.

I put the iPhone back in its display holster and called for Natalie to close whatever app she was exploring. As we ventured outside and searched for a cab, I pondered this new mode of communication. There's something wrong, I thought, when a phone becomes a dependency. What can't it do? More specifically, what can't it do *for you*? I was surfing YouTube recently when I came across a clip posted by German researchers who developed an iPhone app that actually drives a car! Press a button to accelerate, press another button to brake. Want to actually steer the car? Just rotate the iPhone in whatever direction you want to go. The app has not been approved and I hope it never is, for I don't think I can stomach watching my daughter fire up her iPhone and back the car out of our driveway while simultaneously playing *Name That Fart*.

You know who's not on my iPod but should be? The Jonas Brothers. That's right, those three New Jersey kids who along with Justin Bieber dominate every issue of *Tiger Beat* and perform in stadium venues for thousands of prepubescent teens. Even though I don't own their music, I became a fan recently when I discovered that they actually *play* their *own* instruments. I didn't think that was possible anymore. Probably because I recently played *Guitar Hero*.

Guitar Hero and its chief competitor *Rock Band* lets kids actually *think* they are playing an instrument. It comes with a plastic guitar that looks exactly like a regular electric guitar except that it has no strings, no frets, and no tuning keys. Other than that, you can barely tell the difference.

Instead the guitar has five different colored buttons. The guitar "player" starts the game by standing in front of the television screen, the *Guitar Hero* version of sheet music. First he or she chooses a song from the *Guitar Hero* library. I chose "Barracuda" from Heart. From there the song begins and one or several of those colored buttons scroll down from the top of the "fret board." At the appropriate time, the player presses the "note" with one hand and the "strum bar" with the other hand. If done correctly, lo and behold, it appears that you are playing solos just as expertly as Nancy Wilson, Heart's lead guitarist.

Maybe I'm just spouting sour grapes because I took lessons on a REAL guitar for six years. I gave it up and it's a decision I regret to this day for I would love to take a break from time to time in my home office and alleviate writer's block by strumming a James Taylor or Paul Simon tune.

Apple didn't have anything to do with *Guitar Hero*. However, it did create *Piano Wizard PREMIER* which, the website states, "can teach anyone from 3 to 103 how to play the piano and read music in minutes."

Wait, there's more. "*Piano Wizard PREMIER* offers a four-step method that's so simple even three-year-olds are playing Beethoven or Billy Joel in minutes, not months."

I can't speak for either of these composers but somehow I think both would be insulted if they knew Beethoven's Ninth Symphony or "Scenes from an Italian Restaurant" could be mastered that easily. And how about great drummers like jazz legend Buddy Rich or Carter Beauford from the Dave Matthews Band? Would they take a liking to *iDrum Virtual Drum Machine*? I love live concerts but I'm not sure I want to attend a show where four band members bound onstage, fire up their Macs and basically do nothing as the computers play music and the audience screams while waving lighters and iPhones in the air.

I'm not saying all kids are lazy. During the Beijing Olympics we were treated to countless stories of Michael Phelps and his daily four-mile swims, or Shawn Johnson, who walked into an Iowa gymnastics facility when she was five and never left. These kids achieved their success through hard work and perseverance, not because they had something like *iBalance Beam* or *Virtual Breast Stroke* to do it for them.

But just once in awhile, I'd like to see a kid do something—ANYTHING!—without the aid of a computer. I was listening to the local news in my car one cold December morning when I heard the anchor say something about Santa's e-mail address. Apparently kids can now e-mail their wish lists to Santa simply by firing up their PCs (or their MacBooks), typing santa@northpole.com in the subject line and telling Santa how good and obedient they have been, even as their parents scream at them to get off the computer.

I nearly drove into a snow bank. By myself. Without an app.

E-mailing Santa? Does this mean kids will no longer line up at the mall to see Santa? That's a holiday tradition as old as the ugly tie that the aforementioned Aunt Gladys knits every year. You know, the one that smells like cigarette smoke when you open the box?

Santa is not supposed to be reachable via e-mail!!! He's at the mall, somewhere between Macy's and Nordstroms. Kids are supposed to wait at least an hour to see him, as opposed to waiting until an e-mail program loads. Furthermore, kids who do take the easy approach better be prepared for Santa to respond to their request with an e-mail of his own—sometime in February:

TO: BILLY
CC: ELVES
SUBJECT: My absence

Dear Billy: I'm sorry I didn't visit your house this year. I hadn't heard from you. But today I was cleaning out my spam folder and, great green gumdrops, what do you think I found? Your list! Oh well, stuff happens. See you next year at the mall.

P.S. If you still want Guitar Hero, I've had one wrapped up and waiting since last Christmas.

SC

I was still in a "what hath technology wrought" funk as 2010 dawned and I found myself on a business trip to Las Vegas, a city that changes identities more often than Joe Lieberman.

Lieberman is the Connecticut senator who never met a political party he didn't like. A Democrat who endorsed Republican John McCain for President in 2008 and now calls himself an Independent just so he never has to pick up a dinner check, Lieberman was last seen meeting with members of the Whig Party. A party spokesman pledged support, providing Lieberman fought for people afflicted with bubonic plague.

I first performed in Las Vegas in 1991; even then the town was undergoing radical change. Cab drivers, my primary source of information for all things Vegas, were bemoaning the loss of the "Mob influence" that dominated the Strip for the past several decades and apparently made the town run like a well-oiled machine. This was due to the fact that anyone disagreeing with the Mob's decisions was tossed into that machine headfirst and never heard from again.

Vegas cab drivers are among the most entertaining people on Earth. Like everybody in Vegas, they came from somewhere else. Unlike everybody else, they never left. That's because all have some speed bump in their life story that landed them in Vegas driving tourists to hotels, strip clubs and convention centers. All you have to do is strike up a conversation and the sordid detail will eventually emerge.

So, how long you been in Vegas?

'Bout twelve years.

And before that?

I lived in San Jose. I was an IT consultant for Sun Microsystems.

What was that like?

It was great man, particularly the six-figure salary and the international travel.

Did they transfer you out here?

No. My wife—my ex-wife—and kids are still there. (PAUSE) Man, that cocaine will mess you up. Stay away from it, pal. (PAUSE) Did you say the Bellagio or the Venetian?

There it was. No need to press any further.

One time I had a cabbie that spent the first part of the ride berating gamblers.

"So you don't gamble?" I asked.

"Nah, it's for suckers. It does nothing for me. Of course when friends come to visit, I might take 'em to the Strip for the night. Maybe I'll throw a few bucks in a machine here and there. But that's just to kill time."

"So that's it?"

"Yeah, nothing else. Course I might make a sports bet every now and then, especially during football season. Makes the game more fun, right?"

"Absolutely," I said, knowing full well where this conversation was headed.

"Baseball can be fun too. I mean, I'm no Pete Rose or nothin' but I do okay. 'Specially if I parlay. I did a six-team parlay last season that paid me about three grand. 'Course it's Vegas, baby. It all goes back eventually. Hey, who do you like in the fifth at Belmont?"

By the time the cab pulled up to the hotel, I was convinced this guy would bet the over-under on a Pee Wee T-ball game.

Every now and then my Vegas cabbie is a retired guy who's driving a cab because he "missed being around people." That may be true but I can think of about ten thousand jobs in Vegas that allow you to meet people without requiring you to shuttle them between strip clubs at 3:30 a.m. Putting on a full suit of armor and standing outside Caesar's Palace in August sounds more appealing.

In June 1991, I didn't care if Saddam Hussein was running the town, all I knew was that my name was on the Bally's marquee at Flamingo Avenue and Las Vegas Boulevard. There it was, in brilliant red letters:

UNLIMITED SHRIMP!!

Greg Schwem

I have been to Vegas more than 100 times since, watching massive casinos and hotels pop up seemingly overnight as they clamor to attract (in order) gamblers, families, conventioneers, twenty-something partygoers with disposable income and, in 2010, anything that would pull the town out of its horrible economic slump, even if it meant comping waterbed-equipped suites to mass murderers.

I used to spend my free time in Vegas gambling but grew tired of this vice for two reasons:

1) I always lose
2) I never win

Now I prefer to spend my hard-earned money on five-dollar beers in casino bars where I can easily survey whatever slice of humanity ambles through the front doors. On the weekend of January 9–10, I was in Vegas to emcee a sales meeting for a company that makes network certification equipment. If you don't know what that means, don't worry. The employees weren't quite sure either, something that is very common among those who work for high-tech firms. Ask them to explain *exactly* what their product does and they will stutter, stammer and pause before replying it has "something to do with the Internet."

I'd like to say that this sales meeting was THE event in Vegas that weekend. But I would be lying. For there were two other gatherings generating buzz, specifically:

1) The Consumer Electronics Show
2) The Adult Video News Awards

Yes, for forty-eight hours the town would be flooded with geeks and porn stars, the most intriguing combination since Michael Jackson married Lisa Marie Presley.

The Consumer Electronics Show, aka "CES," occurs all over town. Its attendees stay everywhere from swanky hotels like Wynn and Palazzo to fleabag motels that require bedding deposits and still have rotary phones in the rooms, much to the chagrin of CES guests who check in expecting Wi-Fi and docking stations for their iPods. Conversely, the Adult Video News awards take place at a single location, the Palms Hotel and Casino, coincidentally the same hotel that was hosting my sales meeting. That's another thing I love about Vegas hotels: they don't discriminate when it comes to hosting conventions. The Anti-Defamation League could share meeting space with a gathering of Al Qaeda terrorists. As long as both groups take advantage of the all-you-can-eat buffet, Vegas doesn't care.

When my taxi rolled up to the Palms, the marquee did not scream "UNLIMITED SHRIMP" but rather, "AVN AWARDS." What exactly did AVN stand for? It was like trying to decipher one of my daughter's text messages.

A Google search for "AVN" from my cell phone netted several choices: Was AVN a UK-based association of accountants? Was it an awards ceremony for persons afflicted with Avascular Necrosis? Was the Australian Vaccination Network in town? I eliminated those options once I reached the reception desk and saw a mural on the rear wall featuring several leather-clad dominatrices. Also, Vegas hotels don't typically have red carpets snaking from the limo drop-off area through the front doors, running the length of the casino and ending at the Pearl Theatre unless something special will be taking place there. Even though I'm sure it's a big deal to be a member of the Australian Vaccination Network, its members probably don't merit red carpet treatment. And they certainly don't resemble the women who were wandering around the hotel on this particular weekend.

I soon learned that the Adult Video News awards ceremony would take place that evening at 9:00. Suddenly I had plans. By that, I don't mean I decided to attend the ceremony itself. It was sold out and I didn't want to attract attention by strolling through the lobby, holding up two fingers and repeating, "Who's got two?" I would be content to sit at the aptly named Fantasy Bar and watch the scene unfold.

At approximately 6:30 p.m., the first invited guests arrived on the red carpet. I had no idea who any of these women were but their identities seemed to be common knowledge among the hundreds of drooling spectators who lined both sides of the path, making it impossible to get from one side of the casino to the other. Men whose resumes no doubt included "extra on *The Sopranos*" accompanied most of the girls. The guys wore black suits, fedoras, black and white boots, and, if they held onto their "dates" any tighter, Krazy Glue or a staple gun would have been necessary. Apparently "porn star escort" is a highly sought-after position.

The girls paused briefly in front of an AVN banner where they were "interviewed" by a guy who I assume is the AVN's equivalent of Ryan Seacrest. I couldn't hear the questions over the ruckus but what exactly do you ask someone who has been nominated in the coveted "Best All-Girl Three-Way Sex Scene" category? As they posed, hundreds of "fans," most in their twenties and all in need of some direction in life, aimed cell phone cameras and shouted greetings far too crude for this book.

The parade continued for more than two hours. Coincidentally my next-door neighbor from back in Chicago was visiting Vegas the same weekend and joined me in the bar, where we ogled and marveled at the accomplishments in the plastic surgery field. Yet, since we both have two daughters apiece, the conversation eventually turned serious. Where, we wondered, did these girls come from? Where would all these cell phone images end up? No doubt, seconds after most of these electronic keepsakes were snapped on the red carpet, the images were being stored in the iPhone app *Photon* and uploaded to countless Facebook and MySpace accounts so their owners could *prove* that yes, they rubbed shoulders with a real live porn star! Just another benefit of technology.

The next day, once the porn stars and their escorts had checked out and returned to wherever it is that porn stars live, I hopped in a cab and headed a few miles south to the Las Vegas Convention Center, where the CES show was winding down. I had no entrance badge, which posed a slight problem since CES is not open to the public. I wasn't worried; most attendees were leaving the show and rushing to catch courtesy buses that would herd them to McCarran International Airport, also known as "The Last Place to See a Slot Machine for a Few Hours." I could simply borrow a badge and wander carefree through the aisles as "Steve Johnson," "Sharip Dipmarthi," "Linda O'Reilly," "Bill Gates," or whomever would be nice enough to part with their identification. It's a well-known fact that trade show security personnel do not look at badge *names*. Any old badge would do.

Still, securing a badge proved about as easy as winning an AVN award. Maybe it's because identity theft has become so rampant in this country. Whomever I approached responded as if I had just offered them crystal meth. They scurried toward the courtesy buses, leaving me to ponder my next move.

Eventually I realized that a badge wasn't necessary. Hint: If you ever want to sneak into a trade show, remember that there are always thick-necked security guards manning the exits. However, nobody is manning the *emergency* exits.

The Consumer Electronics Show is Broadway for cool gadgets and toys. It also provides a glimpse into the future of home entertainment and communications. Leading electronics players such as Sony, Phillips, JVC and Samsung spend millions on floor space, touting everything from DVD players so tiny they could fit inside a belly button, to camcorders with zoom lenses powerful enough to clearly see what your neighbor down the block is having for dinner. Most of this stuff never actually makes it out of the convention center; in other words, nobody will ever actually *purchase* it. I last attended CES in 1994 and, even then, vendors were showing prototypes of stereos, televisions, cars and homes that I still have yet to see. CES is kind of like those school science films I saw in the 1970s—films with narrators who *insisted* we would all be flying around the world in jet packs by 2000.

Nevertheless, each year CES tech aficionados move through the aisles like kids in candy stores, playing with anything that has a power switch. Representatives from exhibiting companies spend the week getting paid to watch football and play video games on flat-screen televisions. On the day I visited, I watched a Sony employee show an attendee how he could change the hair on his "iPet" while playing PlayStation 3. Does that job come with health benefits?

Trying to see CES in one day is like being dropped into the Louvre for fifteen minutes. Every booth seemed to have something smaller, thinner, brighter, faster, smarter and greener than the one I had just visited. Casio, a company I still associate with cheap digital watches, touted an "eco-friendly projector." This puzzled me. Sure the projectors are void of mercury lamps, which I guess is a good thing, but what happens when the projector is no longer useful? Won't it eventually end up in a landfill?

Motorola wanted me to see phones with "social skills." The folks at Motorola feel it is WAY too complicated to have your Facebook, MySpace and Twitter messages in (horrors!) DIFFERENT LOCATIONS so its new phones allow users to see all the messages at once. The phones sort the messages and, I assume, compose replies if you ask.

Social networking was indeed one of the themes dominating CES 2010 but, as far as I could tell, the theme of the convention should have been, "Touch Me in 3D." Come to think of it, that also would have made a good title for the AVN Awards. Everywhere I looked companies invited attendees to touch their products or see just what it would be like to have 3D technology right in your living room.

First I touched. Microsoft was touting *Surface*, a technology that "responds to natural hand gestures and real world objects."

Incidentally, flipping somebody the bird does not qualify as a "natural hand gesture" because I stood over the "Touch Table" and did it repeatedly. Nothing happened although some Microsoft employees appeared concerned.

The Touch Table is just that—a table you can touch. It looks harmless enough; if I didn't know a computer, projector and several cameras were mounted underneath, I would have been inclined to set a Diet Coke on it. Without a coaster.

I have to admit; the Touch Table provided more fun than I have ever had with a table of any kind other than the one in my fraternity house living room which always seemed to contain at least two of the following: a Tequila bottle, six shot glasses, a remote, and a three-foot bong.

First I watched two Asian men stare at a map of Las Vegas and touch different portions. The map danced under their fingers. They would zoom in and out, pan left and right, and move the entire map from one corner of the table to the next. This seemed to satisfy them and they left.

Then it was my turn. A Microsoft employee showed me how a restaurant of the future might look, with food items that appear on-screen and can be ordered simply by touching them. When ready to pay, simply place your credit card on the screen and drag that image of a T-bone steak directly to the card. Splitting a check would never be so easy. For that matter, neither would waitressing.

As cool as the technology was, all I kept thinking was that the days of xeroxing your ass on a copy machine are definitely over.

After touching, I was ready to learn about 3D. Every vendor at the show decided that just sitting in your family room watching TV was valuable entertainment time wasted. TV should not be watched; rather it should be *experienced.* Mel Gibson should appear to be standing directly in front of my recliner; I'd have to duck to avoid a Drew Brees pass; explosions would seem so close I'd be forced to yell, "Kids, incoming—head for the kitchen!"

The Intel booth was abuzz with 3D gadgets. Standing near a sign that read, "YOUR KIDS WON'T THINK WHAT WE'RE DOING IS CRAZY AT ALL," was an Intel employee, with a 3D camera containing what I assumed was some sort of Intel technology. He aimed the camera at a foreign-looking attendee and asked him to dance.

Amazingly, the attendee complied, gyrating for a few moments as the 3D camera recorded his every move.

"Now put on these glasses and watch," said the Intel guy, handing foreign guy a pair of fragile-looking plastic specs.

I looked at the screen. Since I wasn't wearing glasses, I saw a blurred image of somebody with no rhythm. The "dancer," however, squealed with delight as he saw his lack of dancing ability in 3D. MORE REAL AND RIDICULOUS THAN EVER. The man walked away, giggling with his friend. The Intel employee walked the other way, falsely assuming he would soon have a sale. Apparently nobody told him that CES attendees never actually *buy* anything.

I headed for the exit, convinced CES stood for Completely Expendable Stuff when I stopped. What was this? Something called

ZOMM. It was a product designed to prevent you from losing your cell phone!

Didn't I just write about this problem? Hadn't my daughter just inadvertently abandoned her phone? Was ZOMM the answer to her troubles?

ZOMM, I quickly learned, was invented by Laurie Penix, a mother of three with an extremely difficult last name to type correctly (the proximity of the *x* to the *s* on the keyboard is the reason). The company name is actually a sort of acronym, short for Zachary, Olivia and Madison's Mom. In a YouTube video, Laurie said her kids were victims of "Where is my cell phone-itis" and she wondered what to do about it? So, like any Mom with a thorough understanding of Bluetooth technology, she developed a device that clips to a keychain and emits a signal when the phone's owner walks away from the phone itself. The ZOMM doesn't find the phone for you; it merely tells you that your phone is no longer on your person and, if your name is Natalie, you had better start searching all car bumpers in the immediate vicinity. Sadly, I don't think the ZOMM can help if you drop your phone in the toilet. For that, one must return to the Motorola booth and purchase the Barrage, a phone designed to survive for up to thirty minutes in three feet of water.

I don't know what I found more amazing; the ZOMM or the fact that a busy mom found the time to invent it. My wife Sue is so consumed with shuffling our kids to extracurricular activities that she often forgets to tend to her own basic needs. Sleeping, for example.

Here was a parent who, instead of rolling her eyes at technology, chose to deal with it and correct its shortcomings. Kids lose their phone? Hey, it happens to everybody. Now just carry the ZOMM and that problem is eliminated. Of course you are up the creek without a paddle if you lose your ZOMM but that's something that will most likely be addressed at CES 2011.

Instead of leaving CES disgusted, I left inspired. My funk was lifted. Instead of cursing the day that cell phones found kids and vice versa, I found someone who chose to make the marriage better. It took a parent to create something that all these technology companies had overlooked. Sure, the ZOMM couldn't prevent users from shooting photos of porn stars but I liked Mrs. Penix's style. Maybe parents aren't dorks after all! Maybe we had some *good* ideas about what to do with all this stuff. Maybe there is hope for our world. What would that world be like?

And then I saw *Avatar*.

Released just before Christmas '09, James Cameron's sci-fi extravaganza had already made more money than the gross national product of Sweden by the time I got around to seeing it. It was playing on THREE screens at the Palms Casino multiplex. It also should be noted that no AVN-nominated movies were playing there. Considering the money the AVN group was shelling out, I found that odd. Would it have killed the Palms brass to devote *one* theater to hourly showings of *Tunnel Butts*?

I chose the 3D IMAX theatre (might as well go all-out, I thought—or "all-in" since this was Vegas), donned my plastic 3D glasses, bought an industrial-sized drum of popcorn and a keg of Diet Pepsi and, thirty-five dollars lighter, entered the theatre. Within moments I was transformed into the eye-popping world of Pandora, where ponytailed, tropical ocean-faced inhabitants lived in trees and were so stupidly tall that each would go number one in the NBA draft providing they were willing to get tattooed and carry weapons into their respective locker rooms. Cameron's pitch to studio executives was probably quite simple: "Imagine if Pocahontas married Yao Ming while starring in The Blue Man Group!"

I was transfixed. I listened to Sigourney Weaver speak Na'vi. I watched Stephen Lang play the best badass military type since R. Lee Ermey told a recruit in *Full Metal Jacket*, "You're so ugly you could be a modern art masterpiece!"

Cameron's 3D wizardry made insects buzz behind my ears. Humungous flying lizards swooped down from the heavens and appeared to have their sights set on my popcorn. Brilliantly colored flora and fauna waved in the background. The final battle scene left my ears ringing, my eyes blazing, and my senses on overload. I cheered with the rest of the audience as *Avatar* hero Jake Sully abandoned his human body and became a Na'vi.

Pandora was a world unspoiled by technology, a fact not lost on a section of moviegoers who became depressed after seeing it. The movie's release spawned creation of websites such as naviblue.com, where people could discuss the film. Included in the site was a discussion board entitled "Coping With Avatar/ Pandora Withdrawals."

Here are a few posts, complete with misspellings:

>Ive had Avatar withdrawl... It is not fun... trust me, i wanted and still do want to dream about being on pandora and being a Na'vi and living out their life...

>The point behind Avatar should strike home to anyone that has eyes to see what's happening! I believe I live my life is good as possible but that doesn't mean the planet doesn't have serious problems of its own

---

>Totally agree with ni`sugar. Tsyal Makto seriously needs to get a life, or laid.

>I understand how hard it is to want a perfect world and after seeing Avatar, I just wanted that escape!

Although I have no interest in posting on this site, a part of me understood. Hadn't I been longing for a simpler world? One not laden with adapters, power strips, wireless devices, remotes, thumb drives, webcams and apps? A world not teeming with online perverts, distracted drivers, video-game-induced obesity and libelous statements formulated in chat rooms? Maybe these chat room participants would not be so depressed if the Na'vi had returned to Earth and taught us the beauty of their ways. Maybe *Avatar* needs…A SEQUEL.

This was my Laurie Penix moment. I would write it!!

Of course this was before I read that Cameron already was planning a sequel, which would explore the oceans of Pandora and quite possibly include a scene where the Na'vi discover the remains of the *Titanic* and the body of Leonardo DiCaprio.

Still, I was not deterred. Cameron can take all the credit for *Avatar: Part Two*. But before Jake Sully dives into Pandora's crystal blue seas, unspoiled by leaking BP oil rigs and Navy warships, he has some unfinished business to attend to on Earth. Hence the suits in Hollywood need to contemplate my plot for *Avatar: A Few Years Before The Sequel.*

Welcome to Jake Sully's World

It is 2159. Jake Sully has abandoned his Avatar body and returned to Earth. The reason? Neytiri, the Na'vi he hooked up with on Pandora, told him she wanted a career while Jake wanted kids. They parted close friends. After receiving an honorable military discharge—and a new set of legs—he now sells insurance. He is a little older, a little grayer, and has settled into the comfort of marriage. Back on Earth, Sully married a teacher, who quit her job when child number one came along. Now Sully has two daughters, eleven and sixteen. Some days Sully goes to work while other days he toils in his home office, aided by video conferencing and web-based training. The human race is alive but has changed dramatically. Lady Gaga IV is president; she was elected overwhelmingly as a write-in candidate in 2156, the first election where online voting was allowed. In her acceptance speech she thanked her supporters, "whoever you are."

One afternoon Jake's cell phone springs to life. He receives a call not from a disgruntled policyholder but from an international consortium consisting of cryogenically perpetuated Steve Jobs, Google co-founders Larry Page and Sergey Brin, Facebook founder Mark Zuckerberg, and Nancy Pelosi, who is now doing consulting work. Like the consortium that Sully worked for when he visited Pandora, this group also has a charter that allows it to exploit the resources of other planets. Weapons of mass destruction will not be used. Therefore, Dick Cheney will never become a member.

The group wants Jake to return to Pandora. Satellites have discovered that the Na'vi, while still fond of bow and arrows, now use technology but in ways unknown to those on Earth. Humans still cannot exist on Pandora because there is no text messaging. Jake is asked to once again immerse himself into the Na'vi culture, see how technology is being used and report back to the consortium. If the members find it useful, they will charter monthly flights to Pandora, courtesy of Paris Hilton Airlines. If they find it useless, they will hire a lawyer to discover a loophole in that "no weapons of mass destruction" sentence.

Sully is intrigued. He mentions the opportunity to his wife and daughters at the dinner table. His wife pledges her support but silently thinks it is just another component of Jake's mid-life crisis. His daughters didn't hear a word he said.

CUT TO: Sully climbs into a glass-enclosed chamber. Even though he has legs again, they still creak constantly and crave Advil. A technician tells Jake to empty his pockets of every piece of technology he carries on his being. Jake sadly turns over his Blackberry, iPod, Kindle and car remote. He is, in a sense, naked.

CUT TO: The spaceship is circling Pandora. Jake presses his eyes to the window and smiles. Nothing appears to have changed. It is still the same vibrantly colored, unspoiled world that he visited so many years ago: the brilliant green foliage; the waterfalls; the snow-covered mountains; and the wildlife roaming freely. He would have expected at least one Waffle House to dot the landscape by now.

CUT TO: Jake exits the spaceship and immediately hears sweet melodious piano sounds in the distance. He follows the sound, which grows louder. In a clearing he sees a beautiful Na'vi girl, in her mid-20s. Her name is Ni'Pod. As her fingers move madly across the keyboard, Jake is entranced. He notices a notebook computer atop the piano. Ni'pod notices Jake and stops.

NI'POD

Yes sir?

Jake is taken aback. He had not been called "sir" in a long time.

JAKE

What is that you are playing?

NI'POD

It has no title yet. I am still composing.

JAKE

You composed that? What is the laptop doing? Didn't it compose the song for you?

NI'POD

That is not possible.

JAKE

Yes, it is. Back on my planet we have something called *Piano Wizard PREMIER*, created by Apple. The website says it can teach anyone from 3 to 103 how to play the piano and read music in minutes.

NI'POD

That's like weird, you know?

Jake wonders when the Na'Vi learned to talk like people on Earth, inserting "like" and "you know" randomly into sentences. Was he somehow responsible for that when he visited the first time?

NI'POD

You amuse me with your foolishness, sir.

Suddenly, a shrill sound emanates from a device on the girl's waist. It sounds like something from Jake's childhood —a ringing phone. Jake had not heard that sound in years, at least not since everyone on Earth decided to adopt his own personalized ringtones. For the last six weeks, Jake had answered his phone to the theme song from his favorite television show, Ultimate Fighting in 3D.

JAKE

Are you going to answer your iPhone? I had to surrender mine when I came here.

NI'POD

It is a WePhone.

JAKE

Excuse me?

NI'POD reaches for her phone and hands it to Jake for inspection.

NI'POD

I have read about iPhones in history books. On our planet, we strive to develop technology that brings people together.

JAKE

So no fart apps?

NI'POD

I am not familiar with, what did you call it? Fart? There is no fart on Pandora.

Jake looks as if he is ready to load a moving van and set up permanent residence here.

NI'POD

Please. Look at it.

Jake takes the phone and begins scrolling.

JAKE

You have many apps. What is *GarageScout*?

NI'POD

It is an app that helps the user locate a car mechanic that won't charge him for unnecessary work.

JAKE

Cool. And *Armsaway*?

NI'POD

It detects whether someone is carrying a weapon. Would you find use for it on your planet?

JAKE

In public schools, definitely. How about *Tuition Wizard*?

NI'POD

It tells parents EXACTLY how much college will cost and then tells them EXACTLY how they are going to pay for it.

But if the parents do not like the answers, they can download *4ForLife*, an app that lets a four-year-old stay four forever.

JAKE

What's that?

NI'POD

What's what?

JAKE

You have an app on here called *WhatsThat*? What is it?

NI'POD

It's an app that tells the user ahead of time what commercials will be playing in the midst of programs that you are watching with children.

JAKE

It would eliminate those uncomfortable, "Daddy, what is erectile dysfunction?" questions.

NI'POD

What is erectile dysfunction? It is a phrase of which I am not familiar. Is it similar to farting?

The moving truck was now fully loaded.

JAKE

There is still more. How about *FindaFriend*?

NI'POD

It is used most often in school cafeterias. And here is another app popular among parents. It is called *NoUse*.

JAKE

NoUse?

NI'POD

When a child is born, the parent opens the *NoUse* app and waves it over the child's body. In an instant, the app tells what occupation the child will have when he or she reaches adulthood.

JAKE

And why do you want to know ahead of time?

NI'POD

Let's suppose the child will become a writer. I'm not sure what it's like on your planet, but on Pandora, writers will never use calculus. So there will be no need to take calculus courses in school and, as a result, no need for parents to try and assist with math homework they don't understand.

JAKE

But what if the app reveals the child will grow up to be something horrible? A criminal? A paparazzo? On our planet there was even some guy who convinced every law enforcement officer in Colorado that his son was floating away in a homemade balloon just so he could use the publicity to secure a reality show! Turns out the kid was hiding in the attic. What about guys like him?

NI'POD

Eventually their evil ways will be corrected.

FOCUS ON: A gigantic bird-like creature breathing fire swoops down from the heavens. It flies directly into frame and over Jake and Ni'Pod. Audience members will jump in their seats as 3D technology makes it appear the bird is going to eat their popcorn or, at the very least, singe it. But the bird has something in its jaws. It is a middle-aged man holding a digital video camera and screaming hysterically.

SULLY

What was that?

NI'POD

It is our Intolerus. It flies around Pandora looking for parents who are, shall we say, getting out of line.

SULLY

You only have one?

NI'POD

We have several. Occasionally an entire herd of Intolerus are needed at youth soccer games. The parent you saw became too hysterical at a game. He was shouting obscenities at the coach and insisting his son wasn't getting enough playing time. He is being dealt with.

JAKE

How?

NI'POD

He will be transported to Overinvolved Island. He will meet assorted hockey dads, PTA parents, and helicopter moms. They will undergo therapy and eventually see that children, if left alone, will all grow up to do great things. The parents will be allowed to leave the island on Sunday.

JAKE

And do what? Return to their children's games?

NI'POD

We do not play sports or participate in any organized activities on Pandora on Sundays. That is family time. The tribal elders have dubbed it "Chill Out Day."

JAKE

On my planet, kids are known to spend entire Sundays updating their Facebook pages. Do you have Facebook on Pandora?

NI'POD

We do for we believe in the powers of communication and social networking. But, like everything else here, we have safeguards to make sure this technology is not abused.

JAKE

Such as...?

NI'POD

An application called *GetOuttaMyFacebook*.

JAKE

I like it. Please continue.

NI'POD

Pandora residents are limited to two hundred friends because, let's face it, nobody has more than two hundred friends. You are allowed an additional two hundred acquaintances. These are people who you don't know but who appear to post comments that appeal to you. You are not allowed to respond. You merely read.

JAKE

What is the point of not responding?

NI'POD

It prevents you from getting into conversations and revealing too much information to people you do not know.

JAKE

Can you play games like *Mafia Wars* or *Farmville*?

NI'POD

Of course. Games are very popular on our planet. But if you post too many *Mafia Wars* or *Farmville* updates, or if you post too much information that is deemed to be useless, *GetOuttaMyFacebook* takes over your computer. It will purge your e-mail address book, ultimately leaving you "friendless." Your name and personal information will be eliminated from message boards, chat rooms, groups, and lists.

JAKE

In other words, nobody will know your favorite color and you won't know how anybody is celebrating his or her birthday. You won't know what the weather is in Atlanta, or whose kids just scored a goal, or how late a flight is, or why the president's bailout plan is bad (or good) for America, or how overpriced the hot dogs are at the new Yankee Stadium, or why the new *Ironman* movie ROCKS, or what cute thing the cat did, or why the acoustics are so bad at the concert, or who is procrastinating about mowing the lawn?

NI'POD

Precisely.

JAKE

May I hug you?

NI'POD

Of course.

JAKE

Come to Earth with me. I wanted to stay here the first time but things did not work out. Now I have to report on your ways for a group of people who may be out of control with power. You need to rein them in, show them that perhaps we should slow down our technological innovations and not just create things simply because we can. Teach them to look into the future and decide what is truly necessary.

NI'POD

But where will I begin?

JAKE

Might I suggest a keynote speech at CES?

NI'POD

I will go with you. You seem to be a good husband and father. You need guidance. I will try and provide it.

They do not kiss. Instead they fist bump.

CUT TO: A ship lifts off from Pandora. The camera follows as it ascends. Suddenly it stops mid air and begins to descend again as a persistent electronic beep fills the theater. The audience hears Ni'Pod speak offscreen.

NI'POD

I have forgotten my WePhone.

JAKE

Thank God for the ZOMM.

FADE OUT

ACKNOWLEDGEMENTS

This book is based entirely on actual events and occurrences from my life. While some names have been changed and some dialogue has been taken from memories that fade a bit more each day, every story, anecdote and humorous incident is true.

Diving into the world of self-publishing is a daunting task and I would have surely drowned were it not for the following individuals: Joe Durepos at Loyola Press, for consistently reminding me that my best wasn't good enough; Drew Durepos for valuable editing skills; Becky Anderson of Anderson's Bookstore, for pointing me in the right direction; Judine O'Shea for graphic design expertise and her vision for the book's cover; Paul Crisanti for creating my website, shooting the cover photo and always pushing the envelope; Ray Romano for lending his name to the book; and Bret Nicholaus for great advice. Also, thanks to Laura Swalec at PlumThree for invaluable marketing ideas.

Above all, a heartfelt thanks and a big hug to my wife Sue and my daughters Natalie and Amy. I couldn't have done this without your support and your tolerance. I love you all. Oh, and one final message to my daughters: please turn off the *Wii* and do your homework!

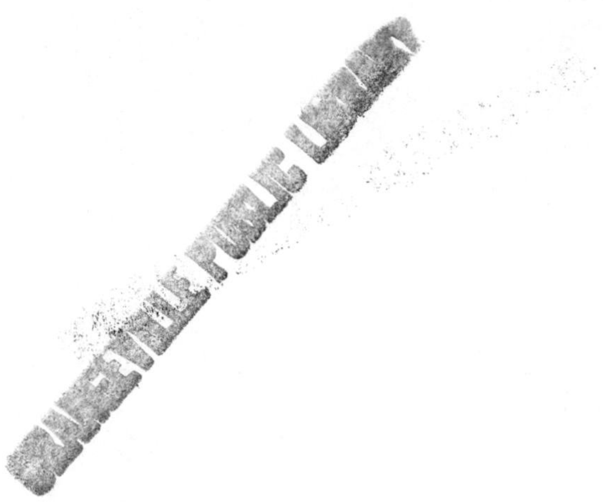